Palgrave Studies in Anthropology of Sustainability

Series Editors
Marc Brightman
Department of Anthropology
University College London
London, UK

Jerome Lewis
Department of Anthropology
University College London
London, UK

Our series aims to bring together research on the social, behavioral, and cultural dimensions of sustainability: on local and global understandings of the concept and on lived practices around the world. It publishes studies which use ethnography to help us understand emerging ways of living, acting, and thinking sustainably. The books in this series also investigate and shed light on the political dynamics of resource governance and various scientific cultures of sustainability.

More information about this series at
http://www.palgrave.com/gp/series/14648

Rosalyn Bold
Editor

Indigenous Perceptions of the End of the World

Creating a Cosmopolitics of Change

Editor
Rosalyn Bold
University College London
London, UK

Palgrave Studies in Anthropology of Sustainability
ISBN 978-3-030-13859-2   ISBN 978-3-030-13860-8   (eBook)
https://doi.org/10.1007/978-3-030-13860-8

Library of Congress Control Number: 2019933314

Cover illustration: © Cecilia Alvarenga/Moment Open/gettyimages

This Palgrave Macmillan imprint is published by the registered company Springer Nature Switzerland AG
The registered company address is: Gewerbestrasse 11, 6330 Cham, Switzerland

ACKNOWLEDGEMENTS

The chapters in this book, with the exception of that by Carolina Comandulli, are the result of the research proceedings of a panel investigating *Climate Change as the End of the World? Mythological Cosmogonies and Imaginaries of Change* at the Royal Anthropological Institute of Great Britain and Ireland's Annual Conference on Anthropology, Weather, and Climate Change held at the British Museum's Clore Centre, May 27–29, 2016. Convening the panel I was delighted by the fascinating array of responses.

I regret that one of the contributions to the panel, *Gods of the Anthropocene* by Bronislaw Szerzinski (published in the journal *Theory, Culture & Society* in 2016), could not be reproduced here, yet I am most grateful that he has been able to contribute an epilogue with commentary on our contributions along with integrated excerpts trimmed from earlier drafts of his seminal article.

I invited Carolina Comandulli to contribute a chapter in the later stages of editing, encountering in her work a future-oriented perspective neatly complimenting the temporality of the other pieces. I am grateful to her for writing a chapter in considerably less time than allowed for our other contributors, and with the additional pressure of a very small son she was caring for.

I would like to thank all the contributors to the book for their hard work. I am delighted with the direction of the eventual work, and have been surprised since the very outset by the fascinating correlations and resonances between the visions of climate change emanating from the various communities we work with.

I would also like to thank Jerome Lewis for inviting me to turn the work of the seminar panel into this book, and to his generous, inspiring, and supportive influence while compiling it at the Centre for the Anthropology of Sustainability, University College London, a center unflinching in its capacity to wrestle with the toughest problems of our time.

I would also like to thank my colleagues Martin Holbraad, Charles Stewart, and Alan Abramson, who contributed to the book's overall orientation as a result of their comments during my time at the department.

My thanks lastly to my father Simon and stepmother Desi, who had me to stay many times at their house in San Roque, Spain, where I edited and collated most of the material. Without the kindness and support of my family I would not have been able to complete the volume.

CONTENTS

Notes on Contributors

Aníbal G. Arregui (Ph.D., University of Barcelona, 2013) has been postdoc researcher at the University of Rostock, the CEFRES—Charles University, in Prague (2018), Beatriu de Pinós (Marie-Curie action) postdoc at the University of Barcelona (2019–present), and lecturer at the University of Vienna (2016–present). Since 2006 Aníbal has conducted ethnography with Lower-Amazonian ribeirinhos and quilombolas, focusing on their bodily responses to environmental, technological, and economic transformations. Since 2018 Aníbal has also been conducting fieldwork in Barcelona, where he follows the proliferation of "urban" wild boars and the eco-political challenges emerging in those urban spaces that are shared by humans and wild species.

Rosalyn Bold gained her Ph.D. from the University of Manchester in 2016, under Penelope Harvey and Peter Wade. She is currently a researcher at the Centre for the Anthropology of Sustainability, University College London. Her research focuses on indigenous perceptions of climate change, contemporary mythological narratives, and the importance of alterity in shaping indigenous politics, with especial reference to Bolivia.

Jean Chamel is postdoctoral fellow at the Muséum National d'Histoire Naturelle in Paris. He holds a Ph.D. in the Study of Religions from the Université de Lausanne. His ethnographic research on European networks of ecologists is related to deep ecology, eco-psychology, eco-spirituality, New Age, eco-centrism, and the rights of nature.

Carolina Comandulli is a Ph.D. candidate at the Anthropology Department at University College London (UCL), funded by the Engineering and Physical Sciences Research Council (EPSRC). She is also a member of the Extreme Citizen Science (ExCiteS) research group and of the Centre for the Anthropology of Sustainability (CAOS). Before starting her Ph.D. she spent four years working for the National Foundation for Indigenous Affairs (FUNAI), which is the Brazilian governmental agency in charge of protecting and promoting indigenous peoples' rights. She also holds an M.Sc. in Anthropology and Ecology of Development from UCL.

Chantelle Murtagh is a Social Anthropologist. She obtained a Ph.D. in Social Anthropology from the University of Manchester in 2016. For her thesis she did research with an indigenous organization in the Peruvian Amazon and focused on indigenous leadership, political organization, and decision-making processes. Previously she has worked as a graduate teaching assistant at the University of Manchester. She has also worked on themes related to indigenous peoples in voluntary isolation in the border region between Peru and Bolivia.

Stefan Permanto was born in 1969 in Gothenburg, Sweden. He earned a Ph.D. in Social Anthropology at the School of Global Studies, University of Gothenburg in 2015. He has conducted extensive ethnographic field research among the Q'eqchi' Maya people in Guatemala and Belize. His research interests include anthropological perspectives on religion, cosmology, rituals, ecology, climate change, migration/integration, as well as issues regarding the rights of indigenous peoples. At the School of Global Studies, Stefan has been lecturing in anthropology, human ecology, Latin American studies, and global studies.

Alessandro Questa is an anthropologist and ethnographer trained in Mexico before receiving his Ph.D. in Anthropology from the University of Virginia (United States). He has worked over the last 14 years with the Mascwal people in the highlands of Puebla, Mexico, publishing on different topics, ranging from migration, ethnicity, and territoriality to material culture, ritual kinship, ceremonial dances and masks, and shamanism. Currently he is writing a book on how Mexican indigenous societies connect their ritual life to local environmental concerns and different forms of activism. He is now Professor Researcher in the Social and Political Science Department at Universidad Iberoamericana in Mexico City.

Bronislaw Szerszynski is Reader in Sociology at Lancaster University, United Kingdom. His research crosses the social and natural sciences, arts and humanities to situate the changing relationship between humans, environment, and technology in the longer perspective of human and planetary history. Recent work has focused on geoengineering, the Anthropocene, and planetary evolution. He is author of *Nature, Technology and the Sacred* (2005), and co-editor of *Risk, Environment and Modernity* (1996), *Re-Ordering Nature: Theology, Society and the New Genetics* (2003), *Nature Performed: Environment, Culture and Performance* (2003), *Technofutures: Transdisciplinary Perspectives on Nature and the Sacred* (2015), and special double issues of *Ecotheology* on "Ecotheology and Postmodernity" (2004), as well as *Theory Culture and Society* on "Changing Climates" (2010—with John Urry). He was also co-organizer of the public art–science events *Between Nature: Explorations in Ecology and Performance* (Lancaster, 2000), *Experimentality* (Lancaster/Manchester/London, 2009–2010), and *Anthropocene Monument* (Toulouse, 2014–2015).

Harry Walker is an Associate Professor at the London School of Economics and Political Science. His current research focuses on ideas and practices of justice in Amazonia and elsewhere, including concepts of fairness and responsibility as well as moral emotions such as compassion and indignation. He is also planning a short film on Urarina ideas about the end of the world.

LIST OF FIGURES

A Territory to Sustain the World(s): From Local Awareness and Practice to the Global Crisis

Relational Ecologists Facing "the End of *a* World": Inner Transition, Ecospirituality, and the Ontological Debate

Introduction: Creating a Cosmopolitics of Climate Change

Rosalyn Bold

The contemporary salience of "the end of the world" as a topic scarcely needs to be highlighted. Social scientists wrestle on the one hand with abundant manifestations of the theme in international popular culture (Danowski and Viveiros de Castro 2017), and on the other address the widespread skepticism that has to date halted effective action on climate change and the environmental crisis (Latour 2017). Speculation on the end of the world is of course coherent with the most careful scientific analyses of the state of things, which show trends hockey-sticking out of the known and into a realm of extreme and irreversible change if we do not halt our consumption, especially of fossil fuels. The Stockholm Resilience Center has identified nine planetary boundaries in a careful attempt to measure damage and chances of recovery in areas of key concern, including the ozone layer, biodiversity, ocean acidification, and climatic change (Rockström et al. 2009); a 2015 update to the Rockström et al. report concluded that four of these boundaries had already been crossed including two key areas, biodiversity and

R. Bold (✉)
Department of Social Anthropology,
University College London, London, UK
e-mail: r.bold@ucl.ac.uk

© The Author(s) 2019   1
R. Bold (ed.), *Indigenous Perceptions of the End of the World*,
Palgrave Studies in Anthropology of Sustainability,
https://doi.org/10.1007/978-3-030-13860-8_1

climate change, which as the report's leading author indicates, "could inadvertently drive the Earth System into a much less hospitable state, damaging efforts to reduce poverty and leading to a deterioration of human wellbeing in many parts of the world, including wealthy countries" (Steffen et al. 2015).

Such analyses too often separate these "natural" factors from the "social," the vast and increasing disparity of wealth itself predicted to topple the world economy sometime soon if no attempt is made to curb the excesses of those at the top (Motesharrei et al. 2014) in the era some call the Capitalocene (Moore 2015, 2017). The Capitalocene, as Brightman and Lewis (2018) indicate, marks where the agency of the Anthropocene is concentrated, in "modern growth based market economies that have intensified resource extraction and consumption around the world, mostly externalising the cost to non-human species and environments" (Brightman and Lewis 2018: 14), yet Anthropocenic thinking can be perceived in all the environments it touches. Separating agentive culture from the mere materiality of "nature", the Anthropocene alienates humans from the environments they inhabit, magnifying *anthropos* against its landscape. Concordantly, the decline in the number of species with whom humans share the earth is so abrupt as to be called the sixth great extinction (Ceballos et al. 2017), a label it seems only contested by those who argue that while we are clearly on the brink of such a change it cannot yet be said to have happened.

Is it too late? James Lovelock, co-creator with Lynn Margulis of the influential Gaia theory (Lovelock and Margulis 1974) in 2008 proclaimed himself of the opinion that it was too late, only to recently conclude in the media that a robot takeover of the planet was likely to be a more pressing concern, introducing a third phalange of eschatological concern, culture/nature/cyborgs.[1] Climate models speak of likelihoods and not of certainties. We are treading into unknown territory; while it is known, for example, that oceans will absorb carbon dioxide from the atmosphere, we cannot know exactly how much nor likewise the impact of flora, expanding in size as the environment becomes richer in carbon, on climatic change.

Out of this uncertainty spring apocalyptic imaginings. For both ISIS and US fundamentalists the end of the world is impending, a situation which neither side dread, as it will evince their status as true believers leading them to paradise. An extremely precarious situation. Perhaps it is, as Pratchett and Gaiman set out in their comic novel *Good Omens* (1990) that despite the darkness of the context we inhabit, the "end of the world" is something we are led into by the narratives we tell ourselves. Turning our attention to innocent pursuits, as the scientist Mayer Hillman indicates,[2] "music and love and education and happiness," which consume no fossil fuels, despite the tangible reality of the four horsemen of famine, pollution, avarice, and war in our everyday lives, we might (still) avoid crisis or, according to Hillman, at least stave it off a little longer.

As Latour indicates, climate change is the "revenge of Gaia" on the modern view in which "nature" is deprived or "de-animated" of agentive capacity, reduced to the status of object for human subject or "culture" to manipulate. Our only chance of escaping it is to think our way out of the modern trap (Latour 2017). Let us therefore construct a cosmopolitics (Stengers 2005) of climate change, widening the concept of politics beyond its modern confines, resisting the tendency of the term to signify, as Latour puts it, "give and take in an exclusive human club"

(Latour 2004: 454) to consult with communities who understand worlds from "other" cosmological standpoints. Cosmos, in Stengers' rendering, refers to the unknown constituted by "multiple divergent worlds, and the articulations of which they could eventually be capable" (Stengers 2005: 995).[3] In this book we encounter various such articulations or reconciliations across registers that are commonly separated as science and mythology.

Mythology is a vehicle specialized in dealing with endings and beginnings and unlike science can construct moral narratives, connecting across the realms of "nature" and "culture." Contemporary Amerindian communities, as Danowski and Viveiros de Castro (2017) indicate, with their sustainable technologies open to complex syncretic assemblages and modest human population sizes, could be considered "one of the possible chances of a subsistence of the future" (Danowski and Viveiros de Castro 2017: 123). In this book we consult such communities on whether the world is ending, whether and why it has ended before, and how we can change contemporary practice to make it sustainable. Can it be saved? What does climate change look like to those who consider that the elements and landscape are imbued with what we might tentatively call consciousness—the capacity to reflect, act, feel? We consider how these communities are equivocating (Viveiros de Castro 2004) their views in cosmopolitical arenas and compare attempts by modern scientists to communicate with them, creating cross-cultural rapprochement.

The Latin American communities consulted here have in common that they inhabit worlds in which humans and spirits co-exist and exchange with one another for their subsistence. Relationships are fundamentally constitutive of the landscapes described, composed of cross-species networks of interacting agentive actors, visible and invisible. In a context where such relationships are more salient than a modern division between nature and culture, changes to the climate are considered inseparable from human actions, and indeed from a sense of crisis subsuming entire landscapes. These are all communities conversant with modernity, most of whom are agentively tackling incursions into their territory, whether from loggers, mining companies, or development agencies, as well as shifting lifestyles and beliefs among themselves.

The communities clearly concur that worlds end when we do not respect the spirits and other species layered into landscapes and indeed within what modernity might term "natural resources." Sharing with other humans and animals that partake in these relational networks is key to entertaining their approval. People are currently ceasing to conduct "earth practices" (Harvey 2007) of respecting and feeding telluric spirits, which across the communities is highlighted as the most prevalent concern. Crucially, we must ask before we take, expressing respect and acknowledgment, what we might call "connection." While they also put forward salient critiques of capitalist extractivist modernity, it may seem remarkable the extent to which these "communities of the future" attribute changes in their environments to their own actions rather than those of *anthropos*; yet, considering that these are landscapes composed of constant conversations, we see how strange it would seem that their actions should be of little significance compared with those of companies and consumers. Climate change for some has the texture of a mood sweeping a landscape, constituted of a network of relationships in which exchange is currently strained. Worlds end, a central theme to all the Amerindian accounts here collected, when people stop engaging in these healthy networks of reciprocity, with both visible and invisible co-habitants, or when relationships are strained beyond their limits. Animist views or relational landscapes are crucial to shaping the sustainable lifestyles, everyday practices, and survival of these "people of the future."

The current moment is widely perceived to be one of crisis. Humans are failing to carry out rituals of respect and acknowledgment for a number of reasons, essentially amounting to the disruption of traditional lifestyles and their accompanying earth practices. Among the Q'eqchi' of Guatemala and Northern Belize (Permanto, chapter "The End of Days: Climate Change, Mythistory, and Cosmological Notions of Regeneration", this volume) civil war has displaced and impoverished communities, who are also struggling to deal with deforestation and depleting animal stocks; in Puebla, Mexico (Questa, chapter "Broken Pillars of the Sky: Masewal Actions and Reflections on Modernity, Spirits, and a Damaged World", this volume) return migrants confront mines, dams, hydroelectric power plants, and gas pipelines locally referred to as *megaproyectos* ("megaprojects") in their territory. In the Andes (Bold, chapter "Contamination, Climate Change, and Cosmopolitical Resonance in Kaata, Bolivia", this volume) the attraction of cities and the cash crop coca, combined with the modern perspective that asserts non-human actors of the landscape are mute "natural resources" for human exploitation, draw the villagers away from fields and ceremonies. In the Amazon the Ashaninka (Comandulli, chapter "A Territory to Sustain the World(s): From Local Awareness and Practice to the

Global Crisis", this volume) combat the incursions of mechanised logging, while the Harakumbut (Murtagh, chapter "Shifting Strategies: The Myth of Wanamei and the Amazon Indigenous REDD+ Programme in Madre de Dios, Peru", this volume) organize themselves against REDD+ schemes quantifying the forest as carbon reserves in a territory suffering from illegal gold mining and overlapped by an oil concession. The Urarina (Walker, chapter "Fragile Time: The Redemptive Force of the Urarina Apocalypse", this volume) inhabit an already fragile world, currently at risk due to the 'weakness' of young people unable to carry out the rigorous shamanic practices to sustain it. Across these communities there is the common conviction that the world is or might be ending if we do not stop and take evasive action. The narratives of the relational worlds described are remarkable for their commonalities, which can furnish an illuminating basis for cosmopolitical dialogue with moderns, some of whom, as we consider in the final chapters (Chamel, chapter "Relational Ecologists Facing 'the End of *a* World': Inner Transition, Eco-spirituality, and the Ontological Debate"; Arreguí, chapter "This Mess Is a "World"! Environmental Diplomats in the Mud of Anthropology", this volume), are already seeking just such "lines of flight" (Deleuze and Guattari 1988).

RECIPROCAL SUBSISTENCE

Firstly, the Amerindian landscapes here encountered all consist of interdependent and mutually sustaining beings. People emphasize the importance of maintaining the environment as a network of beings sustaining one another and requiring human respect.

For the Masewal of Puebla, Mexico, layers of wilful spirits constitute a network of allies and adversaries with whom humans must deal; the landscape is "meshed together by groups sustaining one another" (Questa, chapter "Broken Pillars of the Sky: Masewal Actions and Reflections on Modernity, Spirits, and a Damaged World", this volume). These groups of humans, spirits, and animals are fluid and beings can transform themselves to move between them. Permanto, working with the Q'eqchi' of Guatemala and Belize, expresses similar interdependency as an "existential reciprocity," where humans are dependent on spirits who nurture and care for them yet demand respect. These *tzuultaq'a* oversee animals, crops, forests, humans, and the sky and control the weather

(Permanto, chapter "The End of Days: Climate Change, Mythistory, and Cosmological Notions of Regeneration", this volume). In Kaata, a village in highland Bolivia, human bodies, plants, and mountains mirror one another across scale, ties of reciprocity connecting them; humans "feed" animals, plants, and spirits who "cry" for nurture, and spirits "feed" humans (Bold, chapter "Contamination, Climate Change, and Cosmopolitical Resonance in Kaata, Bolivia", this volume). For the Harakmbut (Murtagh, chapter "Shifting Strategies: The Myth of Wanamei and the Amazon Indigenous REDD+ Programme in Madre de Dios, Peru", this volume) in Amazonia humans are similarly among many agentive actors in the environment, which Murtagh calls, after Surallés and Garcia Hierro (2005), a "relational space." However, there also need to be ordered separations: for the Urarina, Walker emphasizes, animals talking to you might herald the end of *cana cojoanona* ("our days") or a throwback to a previous time in which humans and animals were not yet fully differentiated (Walker, chapter "Fragile Time: The Redemptive Force of the Urarina Apocalypse", this volume).

The Urarina, Walker explains, already inhabit an Anthropocene: in common with the Ashaninka they hold that humans were the first beings, who were transformed into all others. As Comandulli (chapter "A Territory to Sustain the World(s): From Local Awareness and Practice to the Global Crisis", this volume) emphasizes, other beings are therefore still related to the Ashaninka, entailing collaboration and mutual respect. It is crucial that humans ask for permission and negotiate when they need or want something. In other words, as Comandulli expresses it, "what surrounds Ashaninka people are not objects of consumption to satisfy their needs, but a network of relationships that require constant care and attention to be sustained and well balanced" (Comandulli, chapter "A Territory to Sustain the World(s): From Local Awareness and Practice to the Global Crisis", this volume).

Moral Crisis

Considering the spirits in the world around us and treating resources with according respect is crucial and means that the current moment is experienced as *one of moral crisis, concerning human treatment of spirits and others.* Questa explains that among the Masewal remembering spirits is key to sustaining the world:

> The local sense is that an irresponsible embracing of modernity and greed... are the main actions of *amo kilnamitl* ("forgetfulness"), also locally translated as "loss of feeling/emotion", corresponding to an urban and moral wickedness. In contrast, dancing, ascending mountains to give offerings, and unremittingly acknowledging the spirits in the landscape are actions locally associated with *kilnamitl* ("remembrance") a cognate from *ilnamiki* "remembering the dead". (Questa, chapter "Broken Pillars of the Sky: Masewal Actions and Reflections on Modernity, Spirits, and a Damaged World", this volume)

For this population, many of whom migrate to work in Mexico City, "urban wickedness" is associated with forgetting the spirits and evinced by abandoning earth practices associated with place, as well as traditional farming and native languages.

Similarly, failing to uphold the reciprocal bond between humans and spirits, especially the *tzuultaq'a* is for the Q'eqchi' key to the world ending. Should people abandon and eventually forget to conduct the rituals of respect, as elders fear will happen, this will undoubtedly lead to the end of the world. Permanto cites an elder man reporting that when all rituals are forgotten the harvests will yield nothing, animals will die, and people will soon starve. *Tzuultaq'a* are the masters of everything, linked to particular caves, hills, and mountains, multiple expressions of the one *tzuultaq'a*. *Tzuultaq'a* understood in the singular is the mother earth and humanity issued from her womb, a cave:

> Q'eqchi' elders say that they have a strong attachment to their mother earth and that she raises them as her children. They are dependent on her for their survival and sustenance (Permanto 2015: 57) and myths tell of how human infants used to be breastfed by a *Tzuultaq'a* (Cabarrús 1998). Thus, the *Tzuultaq'a* acts as the protecting mother that she is—nurturing and caring for her children, whilst demanding that people are respectful not only to the *Tzuultaq'as* but towards all beings, fellow humans as well as non-humans.

The spirits require considerate and respectful behavior toward themselves, and to other creations with whom we share the earth; they require that humans should "lead a morally correct existence respecting all life, conduct proper feeding rituals, and ask for what they wish" (Permanto, chapter "The End of Days: Climate Change, Mythistory, and Cosmological Notions of Regeneration", this volume).

For the Urarina of the Peruvian Amazon (Walker, chapter "Fragile Time: The Redemptive Force of the Urarina Apocalypse", this volume), sharing is smiled on by Our Creator, whereas being ungenerous or violent risks bringing about the end of the world, envisioned as a cataclysm, when the lightning will "rain down like petrol," land be subsumed by water, or the sky fall. According to one of Walker's informants, "when we eat together, we live like real people. But when our thoughts are different [i.e. when we're stingy], when Our Creator scolds us, the world that he gave us will end." Or as another said: "Some people go around attacking each other, exterminating each other, thinking differently... If we carry on like that we'll all be lost, together with our world. Everyone" (Walker, chapter "Fragile Time: The Redemptive Force of the Urarina Apocalypse", this volume).

Walker explains that *cana cojoanona* ("this world") is maintained by drinkers of the hallucinogenic *Banisteriopsis caapi* preparation ayahuasca, who constantly negotiate for the world's continued survival with the spirit realm. Similarly for the Ashaninka of Amônia River, the visions of their ayahuasca shamans are crucial to shaping their future and surviving the threats they currently confront. Shamans here are community leaders, striving to reverse the destruction they identify in the contemporary world by encouraging widespread reconnection to the sacred. The importance of reciprocity is again highlighted: a shaman, Benki, explains that to exchange is to reconcile differences and bring about reconciliation and transformation. Again the theme of forgetting cultural norms is salient here; Benki criticizes Ashaninka communities for getting so involved with the "outside world" that they are forgetting the "truth" of being an Ashaninka, and calls for them to return to their "own world" (Comandulli, chapter "A Territory to Sustain the World(s): From Local Awareness and Practice to the Global Crisis", this volume).

Among the Harakmbut of the upper Peruvian Amazon the salience of greed in creating the current crisis is highlighted and can be counteracted by communication with the spirit world, as well as by sharing and taking agentive control of territory. Murtagh explains that according to an activist, "our territory or the Amaerikiri reserve is threatened by various activities resulting from man's greed, putting our existence at risk by endangering our water, our biological and cultural resources" (Murtagh, chapter "Shifting Strategies: The Myth of Wanamei and the Amazon Indigenous REDD+ Programme in Madre de Dios, Peru", this volume).

Murtagh records, citing Gray (1997: 119), that "Sharing, abundance and generosity are looked upon favourably and care should be taken not to over-exploit natural resources."

In Kaata, greed is identified as central to the intensification of mining and agriculture, motivating the use of chemical fertilizers and pesticides. These activities strain relationships with the animate landscape through humans "taking too much", thought to bring about such cataclysms as are already occurring in more "developed" countries. Traditional agriculture accompanied by "earth practices" (Harvey 2007) is conceived as "feeding" the land. "Nutrition" is conceived as the product of these healthy exchange relationships; industrially produced foods by comparison convey little nutrition, contaminating bodies and the landscape with their waste, and are weak in relationships. We are at a moment when a mood of contamination sweeps the landscape, weakening the relational network and the beings that compose it (Bold, chapter "Contamination, Climate Change, and Cosmopolitical Resonance in Kaata, Bolivia", this volume).

Greed and sharing with the other humans and non-humans that compose our landscapes are thus identified across communities as crucial to climatic crisis, integrating the realms of nature and society into meaningful moral narratives. Indeed, to cut down on greedy and wasteful behavior and to share among humans and non-humans would undoubtedly help us to survive the combined societal crises of inequality and "natural" crises of environment we confront, even from a modern perspective. In this way the moral mythologies here encountered can illuminate clear axes of change. Relational landscapes can indeed comprise a fitting 'line of flight' from modernity.

We might wonder about the role of Christianity in cultivating these moral understandings. These native communities all identify themselves as Christian, yet manifest very diverse versions of this religion. Andeans, for example, combine worship of elemental spirits and Christianity to engage in what ethnographers have identified as truly heretical practices in small country churches. The heresy of course is in ascribing animacy to the elements and the earth, and dealing with or "worshipping" these actors in and of themselves. Late Christianity renders "nature" God's creation. All the worlds we engage with in this volume are syncretic, yet are so intrinsically connected to the landscape with whom their practitioners are exchanging that this connection seems the more relevant commonality across them.

CLIMATE

Climate change is occurring simultaneously across all these sites, indeed worldwide. While the scale of climate change is terrifying, it is also something we all encounter in common, and it seems vital that we take this opportunity for cosmopolitical conversation. Indeed, it may be that modern conceptions of scale as well as accompanying ontological convictions are complicating the issue, as we shall see in the subsequent section.

In the worlds we consider here climate change is inseparable from other changes; the Masewal, concerned at droughts followed by violent storms, unexpected hailstorms, and massive hurricanes, consider that "[s]tate corrupted agencies, urban vices, mining corporations, and disastrous changes in weather patterns are all elements of the same cosmic cataclysm" (Questa, chapter "Broken Pillars of the Sky: Masewal Actions and Reflections on Modernity, Spirits, and a Damaged World", this volume) caused by human forgetfulness. Yet let us slice momentarily a modern line into these relational worlds and consider changes to the weather in particular, in the hope of opening cross-cultural conversations on that stalwart theme.

The Q'eqchi' note that the weather is becoming increasingly unpredictable, which along with the disappearance of animals is an alarming sign that young people are losing interest in the "old ways," threatening cosmic equilibrium. Weather is sent by the *tzuultaq'a*, to whom people pray for propitious conditions before planting crops. Changes in the weather are making it increasingly difficult to decide when to plant crops, with damaging effects on harvests: "The elders say that it is getting hotter and drier and fear that the landscape will soon become a desert. They no longer hear the thunder as often as they used to and there is less wind and rain, and in some places the coolness of winter is prolonged" (Permanto, chapter "The End of Days: Climate Change, Mythistory, and Cosmological Notions of Regeneration", this volume).

Comandulli reports that the Ashaninka are increasingly concerned about irregularities in the climate directly affecting their livelihoods, as they can no longer trust "bio-indicators"—the behavior of animals, plants, celestial bodies, and weather, which signal when to plant, harvest, fish, hunt, and forage. Similarly, Walker reports, "My friend Lucho would often complain that when he was younger the seasons were regular and one could rely on certain ecological processes to mark out a calendar of sorts: certain flowers and fruits could be expected to appear at the same time each year; and so on. But this was no longer the case, for the seasons are growing increasingly chaotic" (Walker, chapter "Fragile Time: The Redemptive Force of the Urarina Apocalypse", this volume).

In Kaata, crops are said to be "moving up the mountain" as temperatures increase. Villagers take advantage of the different ecological niches the steep sides of the mountain provide to cultivate a variety of crops, with potatoes at the top where it is coldest and maize on the lower slopes where it is warmer. In this landscape such changes are adapted to with fluidity; people now cultivate more maize, while the upper level at which potatoes are grown is continually increasing. It is interesting to note that it is the crops themselves that are said to be moving, rather than humans who are shifting their planting patterns. It seems that this animistic view attributing agency to crops assists adaptation. As Ingold (2006) points out, animists are not "surprised" by changes in their environment; surprise is the reaction of someone who assumes that the "natural" world is unchanging or indeed that we humans have a degree of control over it.

Rituals involving ancestral mummies were until recently carried out to stop an excess of rain in Kaata; these however can no longer be performed as the mummies known as *chullpas*, ancestors of another age of the world who were crucial to the ceremonies, were sold several years ago. It is entirely coherent with this perspective that one of the most damaging effects of climate change is the arrival of sudden intense rains late in the season, inundating seedlings. Cultural change is accompanying weather change across our field sites, which makes sense to their inhabitants.

Walker considers the "epochal quality" of weather: in Spanish or Urarina, as in many of the world's languages, the same word signifies both weather and time, "pointing of course to an intrinsic connection between these two phenomena that is largely lost to those of us living lives largely sheltered from both the fluctuations of the weather and its importance in marking the passing of time." Indeed, *tiempo*, Spanish for weather/time, derives from the Latin *tempus*, as do, as Ingold (2012: 76) indicates, *temper* and *temperament*, describing human moods, thus connecting "our affective lives and the aerial medium in which these lives are led." Humans sustain this weather/time for the Urarina, much like the monks of the temples of the Classical world in Europe. Walker thus raises the importance of sustaining worlds with assiduous temporal care, bringing us again to the importance of "ritual" or communication with the spirits in keeping the world alive, or indeed what we might call willing it to continue.

The connection between weather and mood rings true for the Andes too where, for Bold, "[c]limate change, borne in the winds, has the texture of a mood, a mood of contamination which sweeps, sickening and weakening the lines of relationships connecting the landscape, borne in the winds, or atmosphere" (Bold, chapter "Contamination, Climate Change, and Cosmopolitical Resonance in Kaata, Bolivia", this volume). Where we can converse with the weather we are involved with a sentient animate set of elements; subsumed in a supra-human consciousness. While we might attempt to influence this humor with music, or implore it with days of ceremonies and ancestral mummies, controlling it is perhaps beyond our capacities. The villagers of Kaata were amused at the idea we might "stop" climate change; it is borne in the wind—here a deity, Ankari, which like the Greek god Hermes, carries the messages of the gods to their destinations.

Past Crises

Danowski and Viveiros de Castro (2017) indicate that for certain Amerindian peoples the world has already ended, given the upheaval of colonization; the population is thought to have diminished by as much as 95% of its pre-conquest size by the start of the seventeenth century:

> [V]eritable end of the world experts, the Maya and all other indigenous peoples of the Americas have a lot to teach us now that we are on the verge of a process in which the planet as a whole will become something like sixteenth century America: a world invaded, razed, wrecked by barbarian foreigners. (Danowski and Viveiros de Castro 2017: 108)

Walker remarks on the effect of this massive extermination on global climate, visible in the Orbis spike, a drop in carbon emissions resulting from the change of land use patterns. We might imagine the sense of cataclysm, and of the worlds thereby ending that accompanied this climatic phenomenon for survivors; several authors in this collection hypothesize that mythological references to past apocalypses shaping the present world mark this colonial encounter. Past crises are connected to current conditions; our interlocutors cite mythistorical world endings and compare them with the current moment to assess whether we are indeed at the end of days.

Rupture is central to the landscape of Puebla, where mountains are the "broken pillars" that once connected earth and sky; humans used to climb them when old and descend rejuvenated. The spiteful water spirit *Awewe*, jealous older brother of Christ, smashed the pillars, bringing the rupture of death into the world, and from one unified world there emerged two incomplete but interdependent ones, connected by the winds and water cycles. According to the Masewal, "there is nothing one can see that hasn't been already broken" (Questa, chapter "Broken Pillars of the Sky: Masewal Actions and Reflections on Modernity, Spirits, and a Damaged World", this volume). In the Amazon, Walker explains, the Urarina claim the world has ended before, when the sky fell. It was restored by ayahuasca shamans who negotiated with the spirit world to sustain it to the present. It is however fragile and holding on by a thread, much like a hammock, our world a tiny island at constant threat of being subsumed by cosmic waters flowing from "our source" above and below. When it rains for days or strong winds blow the Urarina fear the world may be ending.

Colonial era texts recording mythological world endings show us how these ideas have merged with pre-conquest cataclysms. Permanto explains that the Popol Vuh, a pan-Maya mythistorical document likely copied from a pre-colonial codex (Tedlock 1996), records that the peoples of this age of the world, the men of maize, were created to sustain and worship the creator spirits. They succeeded men of wood and of mud of ages before them, each of whom were wiped out by these spirits, deciding that the fruits of their creation were incapable of worshipping them properly, the very reason for which they were created. If we do not sustain the spirits, the myth teaches, the "men of maize" of this age of the world may go the same way as those who preceded them.

In the Andes colonial era texts describe *pachakuti* ("cosmic revolution") in terms of environmental cataclysm; while the term is no longer used, villagers in Kaata foresee an explosion of the mountains into streams of lava, engulfing everything, in visions highly reminiscent of these descriptions. Contemporary mummified *chullpas* and their round houses are the remnants of beings that inhabited another age of the world. This ended in an apocalypse, initiating the start of "Christian" times with the rising of Jesus Christ, the sun, who burnt the *chullpas* alive in their east-facing houses. In this myth, like various others in this collection, climate change signals cosmological change, another fascinating commonality. Indeed it seems, in keeping with the idea of weather as a mood of the world, having an "epochal quality," that cosmological change is in all cases here caught up with a change in the weather. When we consider flood myths, said to be a human universal, we see how widespread this idea is.

Villagers in Kaata find that the sun is now "burning them more than before," a warning of a coming world ending. The Ashaninka similarly observe that the sun is getting hotter; the sun here is a creator deity, Pawa, who made the first Ashaninka people. Originally inhabiting the earth he had to leave or it would be incinerated. Moises, a shaman, claims that Pawa is now returning, displeased with his children's actions. Among the Harakmbut the myth of Wanamei recounts a distant time when the whole earth was consumed by fire, and people sought refuge in a tree. The seed of the tree was given over by a parrot, an emissary between this world and that of the spirits, which sheltered all beings until the floods were passed. The myth is brought by leaders into contemporary conversations surrounding the UN REDD+ initiative, the past cataclysm interpreting a contemporary moment, and the fabled presence of the sacred tree, still standing, used as a means to claim and defend territory.

MOMENT OF CRISIS

Across the indigenous groups contributors have worked with, and their mythistories of cataclysm, there is a consensus that we are especially at risk in this moment of the world ending. In the face of contemporary extractive projects and deforestation, spirits may abandon a landscape altogether.

Whilst for the Masewal the end has already come and the world limped on, like others in Mexico they are experiencing unprecedented change: "the incremental effects of decades of intense rural–urban migration, combined with the advent of mining and the progressive abandonment of traditional cultivation, are all locally perceived as linked processes that have effectively destroyed the land" (Questa, chapter "Broken Pillars of the Sky: Masewal Actions and Reflections on Modernity, Spirits, and a Damaged World", this volume). It is perceived that mining companies are after the hearts of mountains, made of precious metals, and if these are taken away "[t]he sustainers will abandon a poor—heartless—mountain just as the water beings will forsake its depths." The mountain will then become a *kostik* ("yellowy") and *chichinawi* ("rotten") place (Questa, chapter "Broken Pillars of the Sky: Masewal Actions and Reflections on Modernity, Spirits, and a Damaged World", this volume). The mountain will produce no animals or maize, and "no clouds will come to nest in its crests nor rivers spring from its base" (Questa, chapter "Broken Pillars of the Sky: Masewal Actions and Reflections on Modernity, Spirits, and a Damaged World", this volume). Mining projects are considered to be aggressive foreign initiatives that can effectively kill a mountain, which would make the spirits of place abandon it also. As Questa indicates, "Mining is hence understood as an act of war."

The Q'eqchi' are similarly concerned at contemporary deforestation, already causing a lack of animals; "[t]o cut down trees and even entire forests uncontrollably, without the consent of the pertaining *tzuultaq'a*, leads to devastating consequences. A hill with a lot of trees growing on it is a fresh, vital, and powerful *tzuultaq'a*, while a deforested hill or mountain is a disempowered or even dead *tzuultaq'a*" (Permanto, chapter "The End of Days: Climate Change, Mythistory, and Cosmological Notions of Regeneration", this volume). Here as well spirits can leave the landscape when extractivist projects move in, resulting in its subsequent death and infertility.

Generational change has a significant role to play in the world ending. The Q'eqchi' and Urarina claim that when the elders who currently conduct rituals die the world will end. Permanto (chapter "The End of Days: Climate Change, Mythistory, and Cosmological Notions of Regeneration", this volume) cites an old man who claims, "We are awaiting our deaths—this is the end of days." For the Urarina, Walker reports, the venerable old Gustodio embodies a strength which young people are thought to lack. Gustodio seems to be a similar figure to Don Ramon in Kaata, a 96-year-old man, hardworking and strong, embodying the qualities of the "people of before." Sudden change is perceived to have occurred in the lifetimes of these elders; we are in a moment of upheaval.

The Urarina claim that young people lack the self-discipline required to fast and thus receive ayahuasca with a clean body, as is necessary to communicate with the spirits; similarly the Q'ikche' complain that contemporary young people do not carry out rituals, their chief cause of cosmological concern. Displacement and the influence of the "modern traditions of foreign cultures" have interrupted earth practices, to which we can add current poverty, meaning they cannot afford to buy incense, cacao, animal blood, and flowers to feed the spirits, items which we imagine were once produce of the forest. Elders claim the young Kaatans are similarly too absorbed in their radios to attend to the old ceremonies, which replace the ceremonial music drawing the elements into exchange.

HOPE

Across a number of communities old and young people are renewing earth practices; indeed this constitutes a central axis of hope and resistance. Questa reports that the Masewal "are attempting to rekindle their relations with spirits, recovering old practices, words, and celebrations. They are also instrumentally re-purposing these cosmological arguments and characters to enforce a new discourse of no confidence in the so-called *megaproyectos* ("megaprojects") and to demand compensation" (Questa, chapter "Broken Pillars of the Sky: Masewal Actions and Reflections on

Modernity, Spirits, and a Damaged World", this volume). Numerous indigenous organizations have been formed to combat these *megaproyectos*, which despite their promises of wealth and development are "systematically rejected by these organizations on the grounds of the insuperable argument that what this myriad of investors ultimately want is their land and water" (Questa, chapter "Broken Pillars of the Sky: Masewal Actions and Reflections on Modernity, Spirits, and a Damaged World", this volume). Mountains themselves are considered to act against these exploitative practices, sending flood waters and releasing predatory animals, with one mountain even trapping a team of British soldiers, widely assumed to be incognito prospectors for foreign mining companies, within it.

The Masewal people are renewing earth practices at sacred sites including a "notoriously effervescent revitalization" of local dances "that mimic numerous spirits and visualize ideal relations of reciprocity between the land and its people" (Questa, chapter "Broken Pillars of the Sky: Masewal Actions and Reflections on Modernity, Spirits, and a Damaged World", this volume). Young people are taking a renewed interest in the spirits and rituals. Similarly, returning to work the land is "a moral stance, a way to enact respect for the land and the spirits and in so doing to remember them and even to be recalled by them, generating reciprocal respect" (Questa, chapter "Broken Pillars of the Sky: Masewal Actions and Reflections on Modernity, Spirits, and a Damaged World", this volume). These practices constitute a conscious process of remembering for the Masewal.

In southern Belize, Permanto notes, elders from four communities have started performing rites collectively for all the communities of the region, asking the *tzuultaq'a* for propitious weather for their crops. They are determined to keep ritual practices going and to share their knowledge with young people. In this way it may be possible to stem annihilation: "Even if people experience times of extreme misery, they may eventually realize that it is through the correct performance of rituals that humanity will be able to enjoy a good life with bountiful harvests, and will take up these practices again" (Permanto, chapter "The End of Days: Climate Change, Mythistory, and Cosmological Notions of Regeneration", this volume). In Alta Verapaz, Guatemala, elders came together as a collective, *Ch'eqil Poyanam*, in 2005 and negotiated with the local authorities to obtain water tanks and *laminas* ("corrugated tin sheets") for roofs, as well as entering into processes to obtain land titles. While there is no explicit statement of a world to come, Permanto notes that death is often conceived here in terms of renewal; when a maize cob dies it spreads its seeds.

Ayahuasca shamanism is for the Urarina central to avoiding the apocalypse. In the altered states of consciousness the drink inspires, shamans
are able to converse with the spirit realm, asking for propitious weather,
health, the replenishment of animals in the forest, and that the world
should not end. "Ayahuasca drinkers are sometimes referred to as *cana
cojoanona nunera* (literally "our world-era's support") and are widely
seen as moral beacons on a cosmic scale, courageous as well as wise and
persistent, leading a highly disciplined lifestyle in the pursuit of the preservation of life" (Walker, chapter "Fragile Time: The Redemptive Force
of the Urarina Apocalypse", this volume). Such communication has the
effect of stemming a process of world decline widely felt to be ever present and pervasive in this fragile world, which is like everything in this
humid and bio-rich environment subject to an inherent process of decay.
Here the apocalypse is a constant threat, "never entirely overcoming the
possibility of courageous action by humankind to save it" (Walker, chapter "Fragile Time: The Redemptive Force of the Urarina Apocalypse",
this volume).

Comandulli works with two shamans of mixed mestizo heritage.
Combining an Ashaninka shamanic tradition with the ability to effectively communicate with the wider world, they are community leaders acting at local and international levels. Comandulli elaborates how
although present-day shamans are perceived to have less power than
the great *antawiari* of the past, these men are leading concerted community resistance and reclaiming their territory. Confronting severe
environmental challenges, the Ashaninka of Amônia River have succeeded in attaining a land title, and organised to control their territory
and create new ways of living sustainably within it, with a remarkable
degree of success.

The Amônia River community have concentrated traditionally dispersed dwellings into a village, built on reclaimed cattle pastures, allowing
them to better control outsider access to their territory and strengthen
collective organization. They have established a cooperative selling crafts
to ensure economic independence from exploitative relations of debt-pe-
onage, and founded a school, which now provides bilingual education
incorporating combining literary, numeracy and training in traditional
activities. They have agreed collective land use patterns, expelling villagers who wanted to continue practicing timber extraction or selling game.
They have reforested 40 hectares of cattle pasture and set up eco-forestry
systems incorporating trees for fruit and timber, attracting game animals to return. In addition, they have boosted fish stocks by establishing

fishing pools and have restored river turtle populations. They maintain gardens that supply the materials they employ in crafts and building. These are the villagers' own initiatives, some of which have later attracted outside funding. They have furthermore established a centre for training outsiders to diffuse knowledge of the jungle with others, and are sharing their experiences with other indigenous communities.

These are success stories effected by communities identified as the most at risk of climate change from a development perspective. Communities of the future are agentively tackling the present. Cosmopolitical communication under these circumstances is increasingly vital and powerful.

COSMOPOLITICAL NEGOTIATIONS

Arreguí explores the possible spaces of environmental "diplomacy" (Stengers 2005) opened up by the shaman Davi Kopenawa and scientist Antonio Nobre. Worlds converge here over Nobre's Biotic Pump Theory (BPT), which holds that certain trees draw the rain, a claim similarly made by Kopenawa, who expresses his view that plantation trees cannot attract atmospheric moisture in the same way. Nobre is fascinated that Kopenawa was aware that trees draw the rain, a conclusion the former reached scientifically. He was even more fascinated when Kopenawa informed him that this knowledge was imparted by "the forest spirits."

While they are in some ways "saying the same thing" in terms of identifying a relationship between trees and rain, Arreguí highlights that the discourses of these two men also differ in vital respects: Nobre conceptualizes the forest in terms of a mechanistic "pump," whereas Kopenawa refers to trees' ability to "call the rain," an agentive and sentient relationality. The two differ therefore on what they consider a "relation" to be. Such contrasting worlds require environmental diplomacy, an awareness of difference, and controlled forms of ontological "betrayal" as each sacrifices norms of their own means of discourse: Nobre publicly expresses a desire to be able to see the forest like the "spirits," and Kopenawa resorts to the written word, each with the intention of communicating beyond their accustomed audiences. Their discussion of the BPT is an instance of cosmopolitical dialogue, as Arreguí highlights, which has considerable ecopolitical "resonance," a mutual amplification of the two discourses in the face of the wider public (Prigogine and Stengers 1984: 46).

Arreguí's exploration of eco-political diplomacy is well illustrated by the Ashaninka shamans, who themselves engage in international milieux. They criticize "science without consciousness," signifying the conscientious and moral use of technology. Their experience of 'science' is it seems connected to the mechanized logging machines which destroyed their families' worlds, much like those Bessire (2011) memorably describes the Ayoreo speakers of the Paraguayan Chaco encountering. Their critiques are embellished with references to "consciousness" and the "mother earth," communicating with a more widely ranging set of eco-political sensibilities, as well as emanating from their experiences consolidating territory and resisting extractivism on the Amônia River.

Scientific and indigenous perspectives are also equivocated (Viveiros de Castro 2004) in the Andean material. Faced with European Community (EC) technicians encouraging the use of pesticides and chemical fertilizers the villagers of Kaata experimented on small plots of land before concluding that, while in the short term the "chemicals" made crops grow better, in the long term the condition of the soil worsened as a result of killing the insects whose movement through the soil creates "pores" through which it "breathes"—indeed, a concern commonly raised by gardeners in the United Kingdom. Here empirical methodology is harnessed to animistic or indeed totemistic conceptions of the mountain as a body with skin to combat the use of agricultural chemicals. The sum of these cosmopolitical conclusions is then used to refute EC aid workers. Scientific assessments indeed agree not only with the view that in the long term chemical pesticides and fertilizers lead to declining soil fertility, but also with the empirical methodology the villagers employ, raising the question of why in supposedly rational modern worlds we continue the widespread use of these chemicals. Considering the mountain as a being with skin, indeed with pores that can close, the detrimental effects of pesticides are highlighted, whereas within the logic of contemporary agribusiness' maximization of profit, they are all but essential. Cosmopolitical conversation can lead to "resonances" (Prigogine and Stengers 1984) that lead us to challenge accepted practices.

In a conversation about the recent influx of contemporary consumer waste to the community, villagers in Kaata are drawn toward the notion of the "natural" as a way to express their landscape as it used to be to the anthropologist, an adept cosmopolitical maneuver. This selective adoption of the modern dichotomy contrasts the "natural world" where everything is part of a system, thus nothing needs to be separated from it, with the contemporary state of things, a world which includes "useless" and "contaminating" consumer waste. We can compare this expression of newly divided worlds to shaman Davi Kopenawa's discomfort with the Portuguese term *meio-ambiente* for "environment" (literally "half-environment") (Kopenawa and Albert 2013), similarly highlighting his discomfort with the divisions modernity entails. Murtagh's community leaders similarly engage with the UN discourse of carbon credits, unwilling to separate the forest into alienated resources. The Indigenous REDD+ scheme seeks to recognize jungles as entire ecosystems. All of these arguments directly engage with the fibers of ruptured modern landscapes, and seem to indicate where they can be reconnected and made whole again, if moderns are ready to engage in an egalitarian cosmopolitics.

While modernity inherits the universalism of Christianity, holding that it is uniquely correct in its descriptions of nature, things are changing in this moment of crisis. On the other side of these cosmopolitical negotiations, environmentally aware Europeans are coming to seek an exit strategy from modernist ontologies. Chamel (chapter "Relational Ecologists Facing "the End of *a* World": Inner Transition, Eco-spirituality, and the Ontological Debate", this volume) explores how relational ecologists, connected to widespread networks such as the Transition Towns movement, largely composed of educated Europeans including some climate scientists, are reaching toward what Descola (2013) might call analogism, equating internal and external transformation. They acknowledge the catastrophic scientific predictions for the near future, and thus the need for transformation of the world system, which must also be a transformation of the self and body. Drawing on European esotericism and James Lovelock's Gaia theory, they see the world as a "system of systems," a composition of species, unifying the within and without.

We could see in this a way of responding to what H. Moore (2018) has identified as major recent changes in scientific conceptualization of environmental relations, tracing the impact of the discovery that most DNA in the human body is microbial. Human bodies are a series of nested environments of multi-species organisms, themselves situated within environments composed of complex inter-relationalities of multiple species. The microbial environments of our bodies are continually changed as we eat, drink, and inhale the wider environments we find ourselves within; "[t]he environment now means everything from the level of the molecular to that of the biome" (H. Moore 2018: 77). As she remarks, these "complex relationalities that morph into socialities, look perplexingly like indigenous ontologies" (H. Moore 2018: 77).

The fractal symmetry across scales that Moore describes resembles the world of Kaatan divinatory healing, where the human body is cured by analogically curing the pertaining points on the mountain body. Indeed this fractal sense of scale continues to characterize contemporary landscapes in Kaata, with actors embracing climate change as common to human and natural bodies. "Nothing is separate, nothing!" a community leader holds, talking about the damaging effects of pesticides on humans, crops, animals, and the earth. Likewise for the scientists Chamel describes, climate change is an interior as well as exterior phenomenon, moral as well as physical, requiring an inner transformation to realize the necessary transformation of the world. Here the challenges of the collapse of scale that Latour (2017) identifies in the

Anthropocene are reshaped, and indeed resonate with indigenous land-scapes in which climate change occurs simultaneously across scales in a way we might describe as fractal (Bold, chapter "Contamination, Climate Change, and Cosmopolitical Resonance in Kaata, Bolivia", this volume). Faced with the end of the world Chamel's relational ecologists seek transformation within themselves as well as in the world beyond.

The communities consulted here all consider the climate key to cosmological changes connoting changes in human behaviour—a widespread conviction. Expressing an inner in an outer state, connoting humor, mood, and temperament by means of the weather, is a common device in Western literature too. The weather can create chaos and overwhelm the human conditions for existence, as in the flood myths common worldwide. This resemblance across scales from the body and humor to the surrounding world constitutes a moral and emotional connection. Despite systematic education in modern culture or simply its prevalence, asserting that the beliefs of your fathers are erroneous, there remains this resemblance and reciprocity.

Reshaping scientific practices to take into account ontological contestation is the central aim of several of the chapters presented here. We can consider these cosmopolitical coincidences or resonances (Prigogine and Stengers 1984) where worlds collide as fertile ground for dialogue, and where we can change practices as a result. In this way we seek to promote sustainability, which Brightman and Lewis (2018) in the first volume of this series define as "a principle based on the active cultivation of cultural, economic, political and ecological plurality, in order to be more likely to address unpredictability in future." To create a sustainable world, we must respect and cultivate diversity at every level.

NOTES

1. https://www.theguardian.com/environment/2016/sep/30/james-lovelock-interview-by-end-of-century-robots-will-have-taken-over.
2. https://www.theguardian.com/environment/2018/apr/26/were-doomed-mayer-hillman-on-the-climate-reality-no-one-else-will-dare-mention.
3. The chapters in this volume, with the exception of Comandulli's, are the result of the research proceedings of a panel investigating *Climate Change as the End of the World? Mythological Cosmogonies and Imaginaries*

of Change, convened by Rosalyn Bold at the Royal Anthropological Institute of Great Britain and Ireland's Annual Conference, 2016, on Anthropology, Weather and Climate Change, held at the British Museum's Clore Centre, 27–29 May 2016.

REFERENCES

Bessire, Lucas. 2011. "Apocalyptic Futures: The Violent Transformation of Moral Human Life Among Ayoreo-Speaking People of the Paraguayan Gran Chaco." 38 (4): 743–757. https://doi.org/10.1111/j.1548-1425.2011.01334.x.

Brightman, Marc, and Jerome Lewis. 2018. "Introduction. The Anthropology of Sustainability: Beyond Development and Progress." In *The Anthropology of Sustainability*, edited by Marc Brightman and Jerome Lewis, 1–34. London: Palgrave Macmillan.

Ceballos, Gerardo, Paul R. Ehrlich, and Rodolfo Dirzo. 2017. "Biological Annihilation Via the Ongoing Sixth Mass Extinction Signaled by Vertebrate Population Losses and Declines." *Proceedings of the National Academy of Sciences* 114 (30): E6089–E6096. https://doi.org/10.1073/pnas.1704949114.

Danowski, Déborah, and Eduardo Batalha Viveiros de Castro. 2017. *The Ends of the World*. Cambridge: Polity Press.

Deleuze, Gilles, and Félix Guattari. 1988. *A Thousand Plateaus: Capitalism and Schizophrenia*. London: Athlone Press.

Gray, Andrew. 1997. *The Last Shaman: Change in an Amazonian Community*. Oxford: Berghan Books.

Harvey, Penelope. 2007. "Civilising Modern Practices. Response to Isabelle Stengers." 106th Annual Meeting of the American Anthropological Association, Washington, DC, November 28–December 2.

Ingold, Tim. 2006. "Rethinking the Animate, Re-animating Thought." *Ethnos* 71 (1): 9–20. https://doi.org/10.1080/00141840600603111.

Ingold, Tim. 2012. "The Atmosphere." *Chiasmi International* 14: 75–87.

Kopenawa, Davi, and Bruce Albert. 2013. *The Falling Sky: Words of a Yanomami Shaman*. Harvard: Harvard University Press.

Latour, Bruno. 2004. *Politics of Nature: How to Bring the Sciences into Democracy*. Cambridge, MA: Harvard University Press.

Latour, Bruno. 2017. *Facing Gaia: Eight Lectures on the New Climatic Regime*. Cambridge, UK and Medford, MA: Polity Press. https://ebookcentral.proquest.com/lib/duke/detail.action?docID=4926426.

Lovelock, James E., and Lynn Margulis. 1974. "Atmospheric Homeostasis by and for the Biosphere: The Gaia Hypothesis." *Tellus* 26 (1–2): 2–10. https://doi.org/10.3402/tellusa.v26i1-2.9731.

Moore, Henrietta L. 2018. "What Can Sustainability Do for Anthropology?" In *The Anthropology of Sustainability: Beyond Development and Progress*, edited by Marc Brightman and Jerome Lewis, 67–80. London: Palgrave Macmillan.

Moore, Jason W. 2015. *Capitalism in the Web of Life: Ecology and the Accumulation of Capital*. London: Verso.

Moore, Jason W. 2017. "The Capitalocene, Part I: On the Nature and Origins of Our Ecological Crisis." *The Journal of Peasant Studies* 44 (3): 594–630. https://doi.org/10.1080/03066150.2016.1235036.

Motesharrei, Safa, Jorge Rivas, and Eugenia Kalnay. 2014. "Human and Nature Dynamics (HANDY): Modeling Inequality and Use of Resources in the Collapse or Sustainability of Societies." *Ecological Economics* 101: 90–102. https://doi.org/10.1016/j.ecolecon.2014.02.014.

Permanto, Stefan. 2015. "The Elders and the Hills: Animism and Cosmological Re-creaction Among the Q'eqchi' Maya in Chisec, Guatemala." PhD, University of Gothenburg.

Pratchett, Terry, and Neil Gaiman. 1990. *Good Omens: The Nice and Accurate Prophecies of Agnes Nutter, Witch*. UK: Gollancz.

Prigogine, Ilya, and Isabelle Stengers. 1984. *Order Out of Chaos: Man's New Dialogue with Nature*. New York: Bantam Books.

Rockström, J., W. Steffen, K. Noone, Å. Persson, F. S. Chapin, III, E. Lambin, T. M. Lenton, M. Scheffer, C. Folke, H. Schellnhuber, B. Nykvist, C. A. De Wit, T. Hughes, S. van der Leeuw, H. Rodhe, S. Sörlin, P. K. Snyder, R. Costanza, U. Svedin, M. Falkenmark, L. Karlberg, R. W. Corell, V. J. Fabry, J. Hansen, B. Walker, D. Liverman, K. Richardson, P. Crutzen, and J. Foley. 2009. "Planetary Boundaries: Exploring the Safe Operating Space for Humanity." *Ecology and Society* 14 (2): 32 [online]. http://www.ecologyandsociety.org/vol14/iss2/art32/.

Steffen, Will, Katherine Richardson, Johan Rockström, Sarah E. Cornell, Ingo Fetzer, Elena M. Bennett, R. Biggs, Stephen R. Carpenter, Wim de Vries, Cynthia A. de Wit, Carl Folke, Dieter Gerten, Jens Heinke, Georgina M. Mace, Linn M. Persson, Veerabhadran Ramanathan, B. Reyers, and Sverker Sörlin. 2015. "Planetary Boundaries: Guiding Human Development on a Changing Planet." *Science*. https://doi.org/10.1126/science.1259855.

Stengers, Isabelle. 2005. "Introductory Notes on an Ecology of Practice." *Cultural Studies Review* 11 (1): 183–196.

Surallés, Alexandre, and Pedro Garcia Hierro. 2005. "Introduction." In *The Land Within: Indigenous Territory and Perception of the Environment*, edited by Alexandre Surallés and Pedro Garcia Hierro, 8–21. Copenhagen: IWGIA.

Tedlock, Dennis. 1996. *Popol Vuh: The Definitive Edition of the Mayan Book of the Dawn of Life and the Glories of Gods and Kings.* New York: Touchstone/ Simon & Schuster.

Viveiros de Castro, Eduardo. 2004. "Perspectival Anthropology and the Method of Controlled Equivocation." *Tipití: Journal of the Society for the Anthropology of Lowland South America* 2 (1): 3–22.

Broken Pillars of the Sky: Masewal Actions and Reflections on Modernity, Spirits, and a Damaged World

Alessandro Questa

In recent times Mexico has become an open field for massive extractive geo-capitalist projects, specifically mines[1] and gas fracking.[2] The so-called "war on drugs" of the last decade has had devastating impacts on regional economies and resulted in widespread migration across and out of the country; its effects are still unraveling.[3] Additionally, the debacle of corrupt state institutions enforced by neoliberal administrations has put millions of Mexicans out of work (Nápoles and Ordaz 2011; Figueroa et al. 2012). Deregulation, corruption, and economic pressure have attracted transnational corporations due to "favorable conditions" for the exploitation of human labor and natural resources, to a point where many Mexicans feel the country is somehow "broken". This chapter deals not only with the particular experiences of anomie and despair but also with the Masewal people's

A. Questa (✉)
Universidad Iberoamericana, Ciudad de México, México
e-mail: alessandro.questa@ibero.mex

R. Bold (ed.), *Indigenous Perceptions of the End of the World*,
Palgrave Studies in Anthropology of Sustainability,
https://doi.org/10.1007/978-3-030-13860-8_2

29

creativity and resistance in the highlands of Puebla state, Mexico. In particular, it explores how certain cosmopolitical actions, keyed as "ritual" by traditional anthropology, are being carried out by locals as a major strategy to halt environmental degradation, seeking protection and negotiating with different local deities and spirits inscribed in the landscape. Indeed, as Bold pointedly signals in the introduction to this volume (chapter "Introduction: Creating a Cosmopolitics of Climate Change"), for a vast number of communities in the Americas environmental changes have cosmological repercussions and causes. While modernity and its transformations evidence the frailty of a delicate cosmos (Walker, chapter "Fragile Time: The Redemptive Force of the Urarina Apocalypse") and the need to—instead of surrendering to it—resist by expanding and re-imagining relations with spiritual agencies (Permanto, chapter "The End of Days: Climate Change, Mythistory, and Cosmological Notions of Regeneration"; Arreguí, chapter "This Mess Is a "World"! Environmental Diplomats in the Mud of Anthropology").

REMEMBRANCE AND FORGETFULNESS

For the indigenous Masewal people (commonly known as Nahua among scholars), living in the eastern mountainous region of Mexico called the Sierra Madre Oriental in the highlands of Puebla state, the world is today breaking apart due to unwarranted human actions against the different spirits that inhabit the land. The Masewal people do not think of themselves, however, as victims of an invasive Western modernity but as agentive actors who have participated in and created this state of affairs. Indeed, most local Masewal people have been deeply invested in all these transformations and have now also formulated a critique of their own enthusiasm and involvement. Around 30 years ago these transformations appeared to offer the beginning of a new era and the end of historic poverty and starvation. But, instead, they have led to a profound dissatisfaction with urban life and a critical evaluation of those same changes and their negative effects on young people as well as on their lands. As my old friend Anselmo, a local healer and traditional dancer put it, reflecting on the recent past:

> Now, why us, here? We have always grown corn; we kept our small *milpa* gardens. [We are] poor people. Back in the city, we were just doing foul things, bad things. Now some [people] have no jobs, others lose our crops, pigs. We come back from there [the city] and we forget [...] the dance helps us to remember how it used to be before, how it must be, in case we forget again. And again.[4]

Like Anselmo, most Masewal people in Tepetzintla, the village where I conducted my research, are concerned with an accumulation of perceivable changes in weather patterns: long-lasting droughts followed by violent storms, unexpected night hail, and massive hurricanes. The local sense is that an irresponsible embracing of modernity and greed (in the form of emigration, together with the progressive abandonment of traditional farming and the decrease in native language usage) are the main actions of *amo kilnamitl* ("forgetfulness," also locally translated as "loss of feeling/emotion"), corresponding to an urban and moral wickedness. In contrast, dancing, ascending mountains to give offerings, and unremittingly acknowledging the spirits in the landscape are actions locally associated with *kilnamitl* ("remembrance," a cognate from "remembering the dead") or *ilnamiki*. Furthermore, loss of feeling and forgetting are evidence of a bigger problem: the lack of *ilnamiki* ("the willing or unwilling detachment by the living from the ancestral dead and from the invisible non-human spirits of the land"). Remembering, in Masewal terms, has to do with a series of collective actions that re-assess the connections among themselves and with their common ancestors inscribed in the land, a practice manifested in different rituals and festivities with which the Masewal people continually hope to recover key "forgotten" knowledge and to enhance respectful relations with a variety of powerful spirits (*espíritus*) that inhabit the landscape.

Forgetfulness and remembrance regularly appear in conversations about local politics, different venues of collective work, and discrepancies regarding ritual protocols and participation. In general, these two concepts encapsulate the tensions between "the young" and "the old."[5] But, beyond being a way to efficiently signal generational conflict, this binary relation is mostly tied to a general concern with numerous recent catastrophic weather events, poor harvests, and economic depression and the ways in which those events can be explained and reversed. This local diagnosis has to do with cumulative collective misbehavior, expressed as "forgetfulness." In turn, such forgetfulness is defined by careless modernization and abandonment of farming and local engagement with traditional ways of doing collective things, referred to as *la tradición* ("the tradition") or *costumbre* ("custom"). Tradition is also the preferred term by which the Masewal people refer to the series of practices and discourses around the ongoing negotiations with local landscape spirits. In other words, remembrance and tradition are tied to a set of relations with the local landscape, while urban practices and forgetfulness are linked to a dangerous detachment from such relations.

HISTORICALLY LIVING IN A BROKEN WORLD

Settled in small rural towns spread across a rugged landscape the Masewal people, historically devoted to growing maize, currently practice a multifarious range of economic and professional activities. However, they keep for the most part a particularly developed set of historic relations and practices with the local landscape. Through well-documented practices, such as growing maize, traditional forestry, or shamanism, as well as less recognized ones, such as masked dancing, the Masewal people display knowledge of and intervene in territories through relationships encompassing wide precepts of inclusiveness and sociality. In general, the Masewal people's ceremonial efforts are linked to maintaining the health and growth of people, land, fields, and animals, ultimately sustaining also *tlaltikpak* ("the world/land's surface").

According to one of the main Masewal stories (those anthropologically labeled as "origin myths") the world is indeed already broken. It was originally fragmented a long time ago by the envy of an ancient being, sometimes called *Nixikol* ("the Devil"), but usually associated with the local *Awewe* ("Water Elder"). This being was the elder brother of Jesus Christ, and was jealous of him, as he wanted people to praise him instead. And so the spiteful *Awewe* picked up speed and crashed against and tore in half, one by one, the many pillars that connected the sky and the earth. Until that moment people would climb up a pillar when they felt old and sickly only to descend from it as children again. In other words, people did not know death. The crumbling of the pillars did not totally separate sky and earth as some beings can still go between them. It implied, however, the partial segmentation of a combined world and the emergence of two incomplete but nevertheless interdependent ones. More importantly, this cosmic end of the world explicates the existence of the current landscape, as the craggy mountains are but the shattered remains of those ancient pillars.

The envious *Awewe* was eventually vanquished, tricked by others to crash against a pillar fabricated out of clouds, then he buried himself at the bottom of the sea. To this day *Awewe* inhabits the depths of the sea and causes storms and hurricanes with his still-thunderous roars—the thunderstorms that signal the beginning of each rainy season—while the rest of the ancient gods remain in the sky and the people on the surface of the earth.

Deities, spirits, caretakers, and gods form today an accumulation of invisible layers of relations in the Puebla highlands, and most Masewal people see no contradiction in this even though the vast majority of them identify as Catholic. They acknowledge Jesus Christ as the only true God and Savior, while also recognizing him as *Totatsin Tonal* ("Our Dear Father Sun"). The sun and Jesus Christ are the same being ruling supreme over all Creation, along with his mother, and sometimes wife, the Virgin Mary who in her many forms is alternatively called *Tonantsin* ("Our Dear Mother"); commonly identified with the earth, fertility, and subterranean spaces and resources, she is the counterpart to the aerial sun.

This recognition of Jesus and the Virgin Mary does not mean, however, that there are no other, older potent beings inhabiting the land and the sky. Masewal farmers acknowledge tempestuous rains, strong winds, and vibrant clouds as the means by which the *Ejekamej* ("Owners of the Sky") speak. The sky and the earth are not fully disconnected and what happens in one still affects the other. Water and winds are paradigmatic examples of these observable connections. According to Masewal lore, clouds use the land to move through forests and ascend to the sky every morning only to go back to "sleep" every night, forming the thick fog characteristic of the region. If water surfaces from underground, according to Masewal observations, it does so to spring up to the earth's surface and then to the clouds only to come back again in the form of rain and mist. Winds are born in distant mountains and crash upon others, bringing clouds and rain, and are sent by the unfathomable will of aerial entities, one of which is the infamous *Awewe*.

The spirits and regions of the world are also connected through a web of kinship. In fact, all the spirits that inhabit and control the underground waters including *Awewe* himself and his wife and daughters, the *Asiwäwewe* ("water women"), are kindred to the windy *Ejekamej* who fly above, as well as those who dwell in the depths of the mountains, the *Tipekayomej* ("mountain bodies"). These related entities can sometimes prove pernicious, desiring humans as sexual partners and taking their souls. An even more ancient pair of spirits called *Tlaltikpak Tata* and *Tlaltikpak Nana* ("Father Earth" and "Mother Earth") also inhabit the mountain landscape. This ancestral couple is in charge of Jesus Christ's world, taking care of the land and its plants, animals, and peoples. This couple is respectfully called *wewetsin* ("dear grandparents") and the Masewal people appeal to them to take care of their crops and animals. Finally, there are also the wild and lascivious spirits of the forest who punish people for mistreating it and chase after lonely women, such as the *Kwoshiwa* ("forest man") also known in Spanish as *Juan Monte* ("Mountain John"). This particularly dangerous being is identified with *Nixikol*, the Devil, and relishes punishing those who wander too deep into the forest as well as those who cut too many trees or kill too many animals. All these layers of wilful spirits constitute a network of allies and adversaries with whom the Masewal people deal. This network forms in the end the ultimate and sometimes conflictive conglomerate of authorities in the highlands.

"Those, the pillars, well they are the mountains [*cerros*]. What's left of them. That's why we go to leave offerings there," commented Francisco, a local diviner, "so they remember us too."[6] From the shattered remnants of gargantuan pillars mountains were made, every one of them a broken and interrupted structure but also a connector of worlds. That is why mountaintops are still partially associated with the sky spirits. Healers and diviners climb them, especially the highest peaks such as Kosoltepetl, Tonaltepetl, or Chignamasatl, so they can see great distances They also leave offerings to the "angels" or the wind spirits, asking them to carry messages, to carry away storms, and to bring good weather and other things associated with the continued sustenance of *life*.

"Life" in Náhuatl is referred to as *yolia*, a concept that evokes the "heart's beating" which is shared in different degrees by spirits, animals, and plants. But life can only be maintained through continued reciprocal efforts, carried out by specific kinds of actions. This "sustained life" is an important Masewal concept, *tlatikipanolistl*, and locally translated into Spanish as *mantenimiento* ("sustenance"), a form of mutual caring expressed through visitations, food offerings, and respectful speech among individual people, households, saints, animals, and landscape spirits. Working in a *milpa* ("cultivated field") is one of the main methods of producing such co-sustenance, although it is not the only one. Through this profound notion of co-sustenance the world—more precisely *taltikpak* ("the world/land's surface")—is reproduced by groups of beings sustaining one another in many and often unaccounted ways. Such categories of beings are more fluid than a rigorous taxonomy would allow. Members of each group, be they people, animals, or spirits, can temporarily take on different forms and act upon others creating yet more rippling effects and relations. For example, *tipekayomej* ("mountain people") can become animals (snakes, badgers, deer) and eat maize from the planted plots. If the animal is killed, however, some human might fall sick to repay for the attack. Similarly, when a tree is cut down without the proper offerings to the same *tipekayomej* the smoke of its firewood can become toxic to the household who felled it and will require the intervention of healers and diviners to appease the angered party and re-establish good relations. Moreover, the specific ancestor spirits inhabiting each house will take offense if not properly remembered by the living occupants, bringing about further harm or at least anxious speculations. These relations are not only dangerous, however, as all these types of beings are inescapably necessary to one another, linked by trophic chains and by dreaming. For example, animals prey on each other, while plants need water and soil, just as humans require crops, herbs, game, and wood and rocks to build their houses. What ties together these relations for the Masewal people is an encompassing idea of interdependence and mutual sustenance.

All these different entities relate to each other in particular ways and react morally to each other's actions. That is, they can become grateful, respectful, supportive, wilful, capricious, or vindictive toward specific people. Hence, for example, a person can dream of giving a feast for a spirit and in consequence have a good harvest while another person will suffer misfortune (accidents, disease, poverty) as a result of neglecting similar courtesies to that same spirit. Spirits meet the Masewal people in their quotidian practices such as dreaming, building houses, or working the fields, creating a pulsating network of relations in constant tension and expansion. Such a hyper-inhabited world requires carefully planned protocols and auspicious negotiations that occupy much time and effort, directed at guessing spirits' intentions, appeasing them, and convincing them to act or to refrain from acting.

An Inclusive Town

Inclusiveness, as a vehicle for diplomacy with modernity, is possibly the common trait for most the different communities included in this volume (Bold, chapter "Contamination, Climate Change, and Cosmopolitical Resonance in Highland Bolivia"; Arreguí, chapter "This Mess Is a "World"! Environmental Diplomats in the Mud of Anthropology"). In a similar way to the Q'eqchi' of Guatemala and Northern Belize (Permanto, chapter "The End of Days: Climate Change, Mythistory, and Cosmological Notions of Regeneration"), who integrate through ritual both spirits and mestizo populations, for the Masewal people spirits are both powerful and needed allies as well as fearsome enemies behind weather calamities and extractivist threats. In Tepetzintla beings (human or otherwise) are not always necessarily visible. Indeed, most of what happens in the world is regarded as imperceptible and carried out by unseen forces locally referred to in Spanish as *espíritus* ("spirits"). In the Náhuatl language, spoken by the Masewal people, spirits have a wide variety of names according to their origins, location, and other contextual characteristics. Spirits are embedded in the landscape, as well as in people's actions, and their influence is often determinant. Hence, visible and invisible beings, human industry, weather patterns, and climatic cycles are all locally considered as elements of *weyaltepetl* (literally "big mountain-town") referred to in Spanish as *el mismo pueblo* ("the same town"), in other words, a total society centered on mountains as the axis for life. This inclusive social nature (Viveiros de Castro 1998; Descola and Pálsson 1996; Descola 2013) gives membership to more-than-human participants and allows for the constant negotiations between different groups, as they all ultimately share and reproduce the same interdependent and circumambient spaces.

Across the highlands of Puebla the Masewal people consider that some form of social intervention is necessary to facilitate the radical transition between life and death (Lok 1991; Knab 1995; Báez 1999; Lupo 2001; Chamoux 2008; Millán 2010; Questa 2010). Most mortuary rites prepare the recently deceased to join the graveyard as the specular "town" of the dead (Pérez 2014). Death implies the final fusion between people and the land (Pérez 2014). The dead will eventually lose their individual identity, to become renamed and associated with specific animals or places, transforming them effectively into ancestors located in the landscape. The result of this process of fusion is a dense landscape that turns out to be the definitive locus for people's efforts, bodies, memories, and post-mortem futures. Sustenance, remembrance, and forgetfulness are thus local concepts tied together in this context of belonging and attachment to the land.

Amerindian cosmologies, although dissimilar from one another in many ways, do seem to share a particular concern with the linkages between living things entwined in ever-expanding assemblages of interdependence, usually expressed in anthropological terms as *animism* (Durkheim 2008 [1912]; Descola 2013). Animals, plants, and spirits all have dealings with humans at some point, and they regularly cohabitate and co-own mountains, rivers, lakes, and forests. The Masewal people consider such relations of exchange to be reciprocal, as they recognize that good fortune during harvest season, for example, is the outcome of equally good relations with certain local entities or spirits. Conversely, if spirits behave sometimes vindictively or reveal themselves to have an avid predisposition to be dangerous, it is because people have been forgetful, lazy, or disrespectful toward them. In other words, spirits' reactions tend to be a reflection of people's offenses. I remember a few years ago, after a couple of days of incessant rain and the growing fear of massive landslides, my host Florencia observed, "we throw them [the spirits] out, we forget about them, we starve them. Then, they get back at us."[7] Indeed, the landslides of 2007 in the wake of Hurricane Dan, which killed an entire family, and in 2013 after Hurricane Ingrid, are still very present.

Such concerns have become a driving force in recent years for people who want to know what is wrong with the weather, the spirits, and the world in general. A group of senior villagers have rekindled the visitation and offerings to the mountain being every year, gathering around a local *tlamatk* ("diviner"), after a local woman's dreams were decided to be premonitory. She dreamt of a tall, bearded mestizo man, who asked her for food and respect, for he was offended and threatened to bring yet more disasters. "He was starving and all alone, poor thing," the woman said, "that's why he was very sad."[8] After a diviner was called and made the proper inquiries through prayer and divinatory visions, it was established that the bearded mestizo was none other than *Tipewewe*, the owner and caretaker of the mountain and all of its inhabitants.

The flow of nurturing relations between people and spirits is assumed to be accumulative over time. When a Masewal person is very sick and cannot be cured at home or at the local clinic a diviner (*tlamatk*) needs to determine the "real" cause, and people often travel between villages to look for the right *tlamatk*. According to diviners the source of any disease is never found in the body, only its effects. Its origins are usually in unconscious offenses to spirits. The diviner urges the afflicted person and her family to remember recent events that might reveal the nature of the spiritual damage. Eventually, someone remembers a dangerous mountain crossing, an unfavorable conversation using bad words (usually with sexual undertones), getting caught in a fierce storm on the way home, or having an ominous dream. Occasionally, the memories can extend back years. The diviner then becomes an explorer who aims to identify and seek out the offended spiritual party, and engage in either negotiation or coercion. Sometimes, the diviner pleads with the spirit and solicits the sick person's family to repay the offense with food and offerings and collectively ask for forgiveness. On other occasions the diviner is more like a hunter, tracking down the offending spirit and coercing it to release the patient's spirit under threat of punishment. As a diviner once lucidly clarified: "Sometimes one asks permission, one pleads, asks forgiveness. Other times I just tell them to go fuck themselves."[9]

The Masewal story about the breaking of the pillars of the sky is not an "origin myth" as such as it does not exactly deal with "the beginning of the world" or with "the source of life", but with how the condition of the world today came about. The story describes a vicious conflict between formidable entities, a struggle that engulfed all the highlands and its peoples and that physically changed the world. It also clearly states who the driving actors are and the hidden rules that govern their actions and desires. The story is then one of how the world has faced destruction and drastic change and how people have changed with it and endured. Hence, even in a situation of dramatic world rupture and the invention of death, there is no mention of the eventuality of total extermination, similar to the case of the Q'eqchi' (Permanto, chapter "The End of Days: Climate Change, Mythistory, and Cosmological Notions of Regeneration"). In such a highly negotiated cosmology extinction is unimaginable, and where there is no acknowledgment of an origin there cannot truly be an end.

These are the basic tenets of spiritual interdependence operating in the highlands and expressed in oral tradition as well as in dramatic performances such as dances. However, the Masewal people are today facing unprecedented transformations in their territory as climatic disasters are perceived to be taking a heavier toll on lives and crops, while mining enterprises are not only changing the economic and political landscapes but the territory and its geographic formations as well. So, what happens in the case of "bigger" and "newer" events, such as catastrophic weather or capitalistic enterprises like mining? How can any cosmology apprehend the incommensurability of such a non-negotiable situation?

MINES, MOUNTAINS, AND LOCAL RESISTANCE

We were walking home and took a tortuous shortcut up the hill with admittedly a great view. While we stopped for me to catch my breath, my friend Rafael said:

> You remember the old bridge that used to be down by the river? That one was destroyed by big rocks that came down the mountain when the hurricane came. They came rolling and there was just too much water with them. They took it down, crushed it. And for weeks we couldn't cross the river. Why is this? Why now? Some say it is because of what the people in Tetela are doing on the other side of the mountain [a mine]. Or is it because we don't respect the elders, the owners, we don't go and visit them [on the mountain]? I don't remember this happening before, when we did [respect them].[10]

Following Rafael's line of reasoning, climatic change is primarily experienced today in the highlands not as a "natural phenomenon," caused and expressed by meta-human conditions, but as a moral crisis. It is not the material by-product of historic capitalism but a form of divine punishment for breaking ancient accords and disrespectful human behavior. Thus, the Masewal people have engaged a complex process of deliberation and atonement through the opposed concepts of forgetfulness and remembrance. These concepts are being deployed in stories and actions in the landscape. Abandoning traditional farming, taking up urban fashions and vices, and crucially "unlearning" the Náhuatl language is locally referred to as being disrespectful and forgetful. As a countermeasure, certain actions, such as yearly festivities and ceremonial offerings, are linked to respectful attitudes of remembrance, as rites imply continuity with the past (Connerton 1989).

As previously stated, for the Masewal people mountains occupy a salient position as complex living entities. Each mountain is a body, composed of many life-forms. The town where I did most of my doctoral fieldwork is called Tepetzintla, a Spanish variation of *Tipesintl* ("mountain-low-place"). The meaning is, however, elusive, even when its roots seem to be very clear. Locals translate Tepetzintla as "at the foothill" or "at the mountain's base." Other translations, offered by a few of the town's young activists, advocate for "mountain of corncobs." Yet another translation, pushed by diviners and some elders, is "place of many hills." The "secret" name is, however, "mountain's seat" or, more accurately, "mountain's ass." This is not the preferred local translation, although it is grudgingly accepted as the "real" one.

The local mountain has an ass, but it also has a head called *tipekwaxochitl* ("mountain's flowery crown") on one of its numerous peaks, and several caves are its *tipenakas* ("ears"), where diviners and sorcerers leave offerings and talk to it, negotiating favors and repaying debts. This mountain also has a vast and enclosed interior or *tipeijtik* ("stomach"). Tremors are not uncommon in the region and some Masewal friends refer to them as "bowel movements." When in 2013 one such tremor occurred, old Anselmo pedagogically explained that "when earthquakes happen it just means water is stirring under the mountain."[11] Indeed, all mountains are water containers from which clouds, rivers, and mist emerge every morning, only to come down again as rains and more mist every night. Some mountains are also packed with grain, others are supposed to be crammed with chickens, turkeys, pigs, and oxen, and yet others contain cars, computers, and airplanes.

Perhaps the most precious part of a mountain is its *tipeyolotl* ("heart").[12] Somewhere inside every mountain is a core whence all life emanates. It is regarded as "its sustenance" (*mantenimiento* in Spanish or *tlatikipanolistl* in Náhuatl) and as the source for all life in the highlands. In the heart of each mountain live its *Tlatikipanojke* ("Sustainers"). Local people often dream and talk with them. Some people say the Sustainers look like an old couple, a man and woman. Others say they look like two *coyokonej* ("coyote cubs"), another term for mestizo children. Others agree that they do look like children but point out that the crafty Sustainers can in fact take the form of any particular child you think of at that moment.

The mountain on which Tepetzintla sits is called *Chignamasatl* ("Nine Deer") and, according to the elders, this mountain's heart is made of silver. Every mountain has a core of some precious metal, either gold or silver. Mining corporations know this for a fact and this is precisely what they are after. According to Masewal villagers, mining companies collect the mineral wealth that is in fact what forms each mountain's heart. The heart is what sustains life of all sorts, and without it a mountain is "poor." There is no word for "poverty" in Náhuatl, so they use the Spanish word *pobre*. "Poverty" is locally understood not as a lack of money but as an absence of life. Hence a "poor" mountain produces no maize and no animals, no clouds come to nest in its crests nor rivers spring from its base. The Sustainers will abandon a poor—heartless—mountain just as the water beings will forsake its depths. The mountain will then become a *kostik* ("yellowy") and *chichinawi* ("rotten") place.

The incremental effects of decades of intense rural–urban migration, combined with the advent of mining and the progressive abandonment of traditional cultivation, are all locally perceived as linked processes that have effectively destroyed the land. These processes are also major spiritual grievances caused ultimately by collective "forgetfulness," leading potentially to another ending of the world. In a rather desperate strategy to counteract this grim eventuality the Masewal people are doing two things. They are attempting to rekindle their relations with spirits, recovering old practices, words, and celebrations. They are also instrumentally re-purposing these cosmological arguments and characters to enforce a new discourse of no confidence in the so-called *megaproyectos* ("megaprojects") and to demand compensation.

Indeed, in recent years the growing presence of *megaproyectos* (mining operations, hydroelectric power plants, fracking, and gas ducts projects) have met organized indigenous opposition in the highlands, such as the *Red Mexicana de Afectados por la Minería, Tiyat Tlali* ("in defense of our land"), *Tosepan Tiataniske* ("united we will triumph"), and the *Organización Independiente Totonaca* (OIT). The promises of wealth, development, and progress made by engineers, government officials, and corporations' PR teams are systematically rejected by these organizations on the grounds of the insuperable argument that what this myriad of investors ultimately want is their land and water. According to Luz Macías, a local Masewal activist and renowned diviner, these are "tales," fake stories that need to be disputed with available local knowledge or "real stories." Fight fire with fire, as the saying goes.

On a less public level, people are re-visiting and in a way re-appropriating the land, from caves and water springs to mountaintops and rivers. People bring saints, place wooden crosses, and call out to angels and protector spirits, offering them food and music in return for forgiveness and sustenance. There is a notoriously effervescent revitalization of traditional dances that mimic numerous spirits and visualize ideal relations of reciprocity between the land and its people. In everyday conversations in domestic spaces I have witnessed young people's increasing interest in talking about ritual protocols and interpretations of spirits' "real" identities and purposes.

Igniting all of these efforts is the encompassing discourse of *kilnamitl* ("remembrance"). What anthropology would usually label "ritual" the Masewal people in Tepetzintla call "remembering." Such memory is more than an individual process, coming into existence as people appropriately enunciate prayers and visit certain *lieux-de-memoire* (Nora 1989) on the mountain; it is a form of habit-memory (Connerton 1989) rooted in continuous action and reaffirmed knowledge. To remember thus implies in Masewal terms a membership (Halbwachs 1992) constituted through knowing a certain number of litanies, secret names, and specific food and regalia arrangements, and visiting places where powerful spirits can be approached and seduced or coerced. That is, to remember implies to move through space up onto and into the mountain.

Memory is embedded in the landscape and vice versa, hence to remember always entails an awareness and organization that has political ramifications (Feld and Basso 1996; Basso 1996). For the Masewal people these political ramifications are manifested in the direct association between the growing presence of mining projects in the region and the latest disastrous storms and hurricanes. As Santiago, a Masewal friend, once aptly put it, "We go to the mountains to visit the caretakers because we need to remember each of them and where they are, what they do: the mountain, the water spring, and the forest. If they are not happy, they will just leave. But us, where are we going to go?"[13]

Mining projects are not merely environmentally damaging projects but aggressive foreign initiatives that can effectively kill a mountain, and local people are convinced that such an eventuality would end up killing them as well, making them "leave the mountain." Mining is hence understood as an act of war. Perhaps that is why mountains defend themselves and defend the Masewal people as well. Mountains can trap foreign people inside, release predators, and send floodwaters cascading down their slopes. The heart of the mountain has immense riches inside, not just precious minerals. "Everything is there," my friend Ramiro said once, "Everything you can ever want is already there. But if you take too much, you kill the mountain and it kills you in turn, somehow. That is why ambitious people get sick or lost inside a cave, like those *gringos* from the mining company. The mountain trapped them inside that day."

The event that inspired Ramiro's reflection occurred in 2004. The *gringos* in question were in fact a group of six British military speleologists performing an "unofficial" military training exercise and who became trapped in the Cuetzalan Tiger cave near the town of Cuetzalan, around 40 miles east of Tepetzintla.[14] Even though they were unable to climb out, to everyone's surprise the soldiers refused the help offered by Mexican rescue teams and demanded, through the British Embassy, a

British team instead. The request was denied and locals eventually saved the arrogant foreigners, but their presence was taken as an unmistakable sign of a failed secret invasion. The news spread rapidly as local people were not terribly keen on the idea that foreign soldiers were secretly exploring their caves, and their "disrespectful" refusal of local help only added to the grievance. Some theories as to their real purposes there still circulate: whereas most people speculate that the *gringos* were in fact *incognito* prospectors for British gold-mining companies, quite a few Masewal friends think the foreign soldiers were actually looking for uranium. The Masewal people agree that the mountain trapped these *gringos* intentionally and released them to rescue teams only after local diviners made proper offerings asking for their liberation.

For Crispín, a diviner, this is not an unusual story at all. Indeed, stories like this abound in the highlands: tales of proud and greedy foreigners (politicians, rich merchants, hippies, non-Catholic pastors, drug dealers, bandits, miners, and engineers) tricked, dissuaded, or even killed by landslides, floods, pests, and disease sent by the crafty mountain spirits, much to the Masewal people's delight. For the Masewal, much like the way the Ashaninka facing logging have returned to their pre-existing relations of respect (Comandulli, chapter "A Territory to Sustain the World(s): From Local Awareness and Practice to the Global Crisis") or the Q'eqchi' acknowledgment of an "existential reciprocity" with diverse spirits (Permanto, chapter "The End of Days: Climate Change, Mythistory, and Cosmological Notions of Regeneration"), their commitment to an interdependency with different beings inscribed in the landscape takes precedence over the interventions of capitalist extractivism.

Each mountain is a sentient entity and reservoir of life, a container for other bodies and laboratory for novel forms, including cars, skyscrapers, and paradoxically even the very oil drills used to destroy them. Every technological achievement, every tool or "weapon" is originally taken from a mountain. According to the Masewal people, under these mountains-as-factories runs a subterranean network of water, gas, and oil that (along with rainbows) connect the highlands, allowing grain, pigs, money, and game to emerge everywhere. The mountains thus form a network of life and in their hearts lie all the corn kernels that will ever be planted, all the plants that will ever grow, all the animals that will ever roam in its forests, and all the people in the highlands.

From what I gather from my innumerable conversations with young and mature Masewal people from Tepetzintla and the surrounding villages and towns, they conceive themselves as long-term caretakers of the sur-face (not the depths) of the *tlaltikpak* ("land") and, to use their own words, they are its "tenants." Masewal farmers speak of their "partners" and *compadres*, who are the invisible subterranean beings that cultivate and co-grow maize with them. The land's true *itekomej* ("owners") and *chanekej* or *chanchiwanej* ("hosts") are the ancestral non-human beings that animate the circumambient land and that are the force behind most of its continuities and its transformations. The fact that humans are just caretakers of the land and not its owners maintains a certain anxiety among the Masewal people who must recognize that there are pre-existing rules and secret relations that precede them. This Masewal *decentered* notion of humanity is key to understanding how they perceive climate change and how they act upon its local effects, starting always from a notion of inter-dependence and co-production with other life-forms and spaces. Hence, can the Masewal people be said to live in the Anthropocene even when they do not recognize anthropocentric views?

To keep going up the mountain and to grow corn are necessary actions associated with traditional forms of remembrance and are locally considered to be inherent capacities of the person, as old Anselmo said, "we don't know anything, we didn't go to school so we grow maize; people are born knowing that." Other middle-aged men are stubbornly returning from the city to grow their *milpa* gardens in the mountain, even when they have other trades and interests, and taking younger gen-erations with them. While sipping coffee on Anselmo's porch one morn-ing we spotted Macario with his four daughters and their dogs going up the mountain trail and disappearing into the woods carrying tools. "Huh. Why do they all go up [the mountain] now?" I asked—not with-out mordacity—as I knew Macario took pride in having been a fancy construction worker in the city for many years. "Because they have to," crowed old Anselmo, "they know they'll be fucked otherwise. We are all fucked."[15] To work the land is more than a rational economic activ-ity or form of social participation. It is a moral stance, a way to enact respect to the land and the spirits and in so doing to remember them and even to be recalled by them, generating reciprocal respect. Respect and remembrance are the key moral attitudes that sustain life in the high-lands. Mountains, as broken pillars, are the main repositories for these localized actions and stories, providing an alternative and unchallenged relation with time as, for the Masewal people, memory has trapped itself in the circumambient space.

Masewal farmers constantly link the damaging current effects of climatic change to expansionist capitalist enterprises in the highlands and what they consider unhealthy manifestations of modernity (such as urban vice, laziness, poverty, and the abandonment of traditional farming) in a joint moral crisis. The world, already broken, is somehow breaking apart yet again, this time under the pressures of a kind of immoral Anthropocene. By retelling stories, reclaiming old practices, and organizing pan-ethnic groups against ravenous *megaproyectos* the Masewal people attempt to make sense of but also to halt the spreading damage. Political organization and awareness as well as ritual are venues of *remembering* in Masewal terms. Such remembrance activates an ecology in which mountains, and not people, are the centerpiece. From a Masewal perspective, mountains are the true peasants in the highlands, devotedly growing forests, animals, *milpas*, and people.

In other words, by inventing an ecology of interdependence the Masewal people have concluded quite logically that there is a problematic moral content to climatic disasters and that they should be fixed by changing people's disrespectful behavior.

NOTES

1. http://sipse.com/mexico/concesiones-empresas-mineras-trasnacion-ales-mexico-215593.html.
2. http://eleconomista.com.mx/estados/2016/01/10/exploran-fracking-233-pozos-puebla.
3. http://www.nexos.com.mx/?p=27278.
4. Ora ¿Por qué a nosotros aquí? Siempre sembramos el maíz, cuidamos nuestras milpitas. [Somos] Gente pobre. Allá en la ciudad nomás hacíamos cochinadas, cosas malas. Ora ya no tiene trabajo, otros perdemos cosecha, puercos. Venimos de allá [la ciudad] y nos olvidamos [...] la danza es para acordarnos cómo era antes, cómo tiene que ser, por si se nos olvida otra vez. Otra vez (Field Notes 2014).
5. *Telpochme iwan ichpochme* "young men" and "young women", respectively and *weweme iwan tenantsime* "male elders" and "female elders", respectively.
6. "Esos, los pilares, pues son los cerros. Lo que quedó. Por eso dejamos ofrenda allá. Para que nos recuerden también" (Field Notes 2015).
7. "Los sacamos pa' fuera, ni nos acordamos. Los dejamos sin comer. Luego, tienen que venir" (Florencia, Field Notes 2015).
8. "... estaba todo hambreado y solito, pobrecito. Por eso estaba triste" (Florencia, Field Notes 2015).
9. "Bueno, a veces uno pide permiso, ruega, pide perdón, otras sí les digo que se vayan mucho a la chingada" (Anselmo Reyes, Field Notes 2014).
10. Se acuerda del puente Viejo que estaba en el río? Ese fue el que se lo chingaron las piedras cuando el huracán. Se vinieron pa' bajo con harta agua. Lo tiraron, lo chingaron nomás. Y luego pasaron semanas y no se podia cruzar. Porqué pasó eso? Porqué ahorita? Unos andan diciendo que es por lo que andan hacienda del otro lado, por Tetela? O major es que como ya no hay respeto a los de antes, a los dueños, que ya no vamos para arriba [a la montaña]? Yo no me acuerdo que esto pasara antes, cuando sí había [respeto].
11. "Ora, cuando tiembla es nomás por el agua que se está moviendo abajo."
12. Another small riddle in the names, *yolotl* ("heart") and *yoli* ("to be born") are both associated with *yolia* ("life"). Life is animation, beating, and movement. What is born from a woman's womb is a beating heart.
13. "Vamos al monte a visitar a los itekomej, porque hay que recordarlos, dónde están, qué hacen: el cerro, el ojo de agua, el monte. Si no están contentos se van luego. Pero nosotros, ¿a dónde vamos a ir?".
14. http://archivo.eluniversal.com.mx/primera/16159.html.
15. "Ora'. ¿A porqué van todos pa'rriba?" [...] Porque ya saben que si no se van a chingar. Nos chingamos todos" (Field Notes 2013).

REFERENCES

Báez, Lourdes. 1999. *EL juego de alternancias: la vida y la muerte. Rituales del ciclo vital entre los nahuas de la Sierra de Puebla.* México: SEP, ENAH, INAH.

Basso, Keith H. 1996. *Wisdom Sits in Places: Landscape and Language Among the Western Apache.* Albuquerque, NM: University of New Mexico Press.

Cardinale, B.J. 2013. Towards a General Theory of Biodiversity for the Anthropocene. *Elem Science of the Anthropocene* 1 (December 4).

Chamoux, Marie-Noëlle. 2008. "Persona, animacidad, fuerza." In *La noción de vida en Mesoamérica*, coord. by Perig Pitrou, María del Carmen Valverde Valdés, and Johannes Neurath. México: Centro de Estudios Mayas-Instituto de Investigaciones Filológicas-unam/cemca (col. Ediciones especiales, No. 65).

Connerton, Paul. 1989. *How Societies Remember.* Cambridge, UK and New York: Cambridge University Press.

Descola, Philippe. 2013. *Beyond Nature and Culture.* University of Chicago Press.

Descola, Philippe, and Gísli Pálsson (ed.). 1996. *Nature and Society: Anthropological Perspectives.* Taylor & Francis.

Durkheim, Emile. 2008. *The Elementary Forms of the Religious Life [1912].* Oxford University Press.

Feld, Steven, and Keith H. Basso, ed. 1996. *Senses of Place.* Santa Fe, NM: School of American Research Press.

Figueroa, Esther. 2012. "Análisis del desempleo, la migración y la pobreza en México." *REDALYC, Sistema de Información Científica, Red de Revistas Científicas de América Latina y el Caribe, España y Portugal*, Quinta Época. Año XVI. Vol. 30. January–June.

Halbwachs, Maurice. 1992. *On Collective Memory.* University of Chicago Press.

Knab, Timothy J. 1995. *A War of Witches: A Journey into the Underworld of the Contemporary Aztecs.* San Francisco: Harper.

Lok, Rossana. 1991. *Gifts to the Dead and the Living: Forms of Exchange in San Miguel Tzinacapan, Sierra Norte de Puebla, Mexico.* Leiden: Center of Non-Western Studies of the Leiden University.

Lupo, Alessandro. 2001. "La cosmovisión de los nahuas de la sierra de Puebla." In *Cosmovisión, ritual e identidad de los pueblos indígenas de México*, coord. by Broda y Báez-Jorge. México: Conaculta/FCE.

Millán, Saúl. 2010. "La comida y la vida ceremonial entre los nahuas de la Sierra Norte de Puebla." *Diario de Campo*, No. 1, Nueva Epoca. México: INAH.

Nápoles, Pablo, and Juan Luis Ordaz. 2011. "Evolución reciente del empleo y el desempleo en México." *Economía* 8 (23) (May–August): 91–105. http://www.revistas.unam.mx/index.php/ecu/article/view/44995.

Nora, Pierre. 1989. "Between Memory and History: Les Lieux de Mémoire." Special Issue: Memory and Counter-Memory, *Representations* 26 (Spring): 7–24.

Pérez, Iván. 2014. *El inframundo nahua a través de su narrativa*. Etnografía de los Pueblos Indígenas de México. México: Instituto Nacional de Antropología e Historia.

Questa, Alessandro. 2010. "Cambio de vista, cambio de rostro. Parentesco ritual con entidades no humanas entre los Nahua de Tepetzintla, Puebla." Tesis de Maestría en Ciencias Antropológicas, Instituto de Investigaciones Antropológicas, Universidad nacional Autónoma de México.

Viveiros De Castro, Eduardo. 1998. Cosmological Deixis and Amerindian Perspectivism. *Journal of the Royal Anthropological Institute* 469–488.

Fragile Time: The Redemptive Force of the Urarina Apocalypse

Harry Walker

INTRODUCTION

That the world as we know it is on the brink of collapse is a recurring theme among the Urarina people of Amazonian Peru. In this, of course, they are far from alone: generalized anxiety about the future of human-kind might well be almost as old as humankind itself, but seems today on the rise as never before. Apocalyptic thinking is certainly prevalent among the native peoples of lowland South America where, as more than one commentator has pointed out, it has a solid historical precedent, insofar as the world already ended at least once before with the arrival of Europeans some five centuries ago (e.g., Bessire 2011; Danowski and Viveiros de Castro 2017: 104; Neves Marques 2015). Between 1492 and 1610 some 95% of Amerindians are believed to have died: a mass extermination so catastrophic that the resulting shift in land use patterns led to a dramatic global drop in carbon dioxide emissions, the so-called Orbis Spike that has been proposed as a good candidate for dating the start of the Anthropocene (Lewis and Masin 2015). At this same time,

H. Walker (✉)
London School of Economics and Political Science, London, UK
e-mail: h.l.walker@lse.ac.uk

© The Author(s) 2019

R. Bold (ed.), *Indigenous Perceptions of the End of the World*,
Palgrave Studies in Anthropology of Sustainability,
https://doi.org/10.1007/978-3-030-13860-8_3

of course, trade became truly global as the two hemispheres were connected, the New World meeting the Old in a geologically unprecedented homogenization of the Earth's biota (Lewis and Maslin 2015: 175).

For those native peoples of the Americas whose lives were most directly affected, however, the local ecology was already highly anthropogenic; so far as they were concerned, they had always lived in the Anthropocene. That is to say, human action was never really decoupled from atmospheric, geologic, and hydrologic processes; nature and culture, or the cosmological and the anthropological, were never separate. This is of some relevance to us here today, as we finally draw a similar conclusion, right about the same time we are confronted with own narratives of impending disaster. The fragility of the climate, its surprising dependence on human activity, and its shifting nature as a moral and social as well as a technical problem are just some of the areas of common ground that seem worthy of further exploration.

What is of interest for us here, as we struggle to think through the challenges that surround effective governance of the global commons, is more than the Urarina people's careful custodianship of their planetary ecology. I want to draw attention to their acute sense of the weather or climate itself as a common good to be collectively maintained, even collectively produced. In exploring in this chapter how the Urarina are thinking about the imminent end of days, I hope therefore to provoke reflection on the very concept of weather and its relationship to time, and how these together implicate particular forms of agency and responsibility. I wish to show how, according to Urarina apocalyptic discourse, catastrophic climate change is a moral and social as well as technical problem; one over which we do have some influence, even if it is never something we can fully hope to master. When it comes to the apocalyptic we are always in the presence of the mysterious and the unveiled. This is surely something we should never lose sight of if only because it is part of the way in which eschatology, or the theory of final events, imbues life with a sense of structure and purpose.

At the end of the day, however, for all the common ground—our common predicament, as it were—we should bear in mind that there is more than one end of the world: that is to say, the ends of the world are many. There is something quite specific about the end of the world as it is imagined by the Urarina people, and revealing this means also coming to terms with Urarina visions of the world as potentially shaped by, and in turn helping to shape, a specific but diverse range of social, historical, and material factors, from the often turbulent nature of contact with outsiders, to their particular conceptions of the good life, to the ways in which people experience the everyday ebbs and flows of their local riverine environment. And it means paying careful attention to what they have to say about the time we have remaining.

TIMES OF FRAGILITY AND DECAY

I met Custodio at the end of 2005. He was in many ways the last of his generation, living a modest but highly mobile existence with his youngest wife and small children, now too old to hunt or work in the garden but still able to eke out an existence and maintain his independence. He died just a few months after I met him, and his passing was acknowledged by virtually everyone in the area as a significant event, given his prodigious wisdom and his steadfast dedication to the ways of the ancients. Unlike his more sedentary children and grandchildren he still spent his life on the move, constantly traveling and visiting his many descendants, unwilling or unable to stay still for long. He expressed eloquently the lingering dread that I soon learned many others also felt, concerning a near future rushing headlong toward us, a future that nevertheless recalls and in some ways repeats a distant past. His words are well worth quoting in full:

> A long time ago, it's said that our land was destroyed. The sky fell, it's said. It fell to the level of the forest. Everything was destroyed. The ancient people drank ayahuasca, and in that way managed to detain it, up until today. That's why we are afraid. And so we should be afraid! Each month, when the moon is full, it might fall down, finishing us off. How huge must it be? It would completely destroy us! It would destroy the land, destroy our epoch. Then we'd all be completely lost. Otherwise, a tremendous earthquake will cause the land to fly. It will fly away immediately, as though it were made of cotton! And so we'll all be lost. We'll no longer be like people. Otherwise, it will grow dark. We'll lose our marker, the sun. When the sun disappears, there'll be nothing but darkness. Nothing even to make a fire with. It'll be as though our eyes were shut. Jesus Mary, poor little God! Otherwise, the thunder will make a tremendous noise and the lightning will piss down on us from the sky, burning everything, like gasoline it will burn! Boom! Here and there, all around us, the land will sink immediately. Everything that lives will suddenly perish. That's what the ancients said. So yes, please, be afraid, God bless. God who created us. In this way I beg God to bless me. We share food with everyone, don't be stingy, that's what the ancient people said, or God will punish us. Eating together, we can beg God to let the animals appear in the forest. If he turns his back on us there will be no animals whatsoever for us to eat. A great fire will come. "Where is God now?" they will ask, burning in the fire.

All the Urarina I spoke to about the end of the world assented to this narrative, or something roughly similar. There are several points about it worth noting. First, the end of the world is marked by a rapid deterioration in the weather, including a number of elemental forces that pose an immediate danger to life on earth, including the possibility that the moon will fall out of the sky. Second, what is actually about to end is *cana cojoanona* (literally "our days"), which I translated as "our epoch"—suggesting a cyclical calendric of sorts. Third, the end is apparently linked to a decline in good behavior, suggesting an important connection between the earth's climate and the moral order. Finally, the world in fact already ended, a long time ago—although it was effectively restored and stabilized by diligent drinkers of ayahuasca, thus carving out an important space for human agency.

To take the last issue first: what should we make of the apparently paradoxical assertion that the end of the world has already taken place, that the sky has already fallen, sometime in the distant past, and this is why the world remains so fragile today? As noted above, it is tempting to point to the obvious historical correlate for an end of the world that already happened: as Harkin (2012: 97) put it, "For Native Americans in the wake of European contact, the changes were so extreme that they were viewed on the level of the mythical events that led to the creation of the world itself." There is, however, also a certain millenarian element to this discourse, which seems inflected in some way with Christianity, with varying degrees of explicitness. Consider the following story about how the Urarina people were abandoned by God long ago:

> So much did we annoy our Father that he passed over to the sky. With a cross he went, so it's said. And everything over here in our world began to break down. Even today, it's still broken, our world is going to end. It's on its way. That's why, they say, the animals too want to return [to the sky], they want to go back, because it's like this. The mestizos also say this. It's made it up to now, but the world will be annihilated. When the mestizos knew this, they said it too. Yes, they said, it's going to end. All our world, our land. The ancients knew. It will be extremely dangerous. The demons, the unknown ones, are already talking.

Although there were virtually no functioning Urarina churches at the time of my fieldwork and very few if any Urarina could really be considered Christian, it is worth emphasizing that they have had a relatively long history of exposure to Christian ideas and influences. They were first contacted by Jesuit missionaries in the seventeenth century, and for part of the eighteenth century several hundred Urarina lived in the mission settlement of San Xavier de Urarinas, founded on the banks of the Chambira River. The Jesuits were expelled from Peru in 1767, and the Urarina have probably had intermittent contact with Christianity ever since. Christian, especially Catholic, themes are discernible in some of their most well-known myths, which tend to superimpose newer, Christian characters and motifs over pre-existing ones, while recognizably Christian images have been incorporated into ayahuasca-based shamanic chanting. Even shamanic visions blend a recognizably Christian apocalyptic with recognizably Amazonian themes.

It is also true that apocalypticism of the kind we are dealing with seems often to be a discourse employed by subaltern peoples with a history of subordination or exposure to violence. In the words of Fenn (2003: 110): "At times of crisis, when a way of life seems threatened with extinction, the apocalyptic imagination is liable to flare up with special force. In these circumstances, predictions of an imminent end may take on new plausibility." Yet memories of the past are also re-interpreted and re-shaped in light of more recent events, especially those that tend to provoke anxiety and uncertainty. Among the Duna people of Papua New Guinea, according to Nicole Haley, the apocalyptic vision "problematizes the space development inhabits, through its claim that mining will bring about the end of the world … [and] allows Duna men and women to comment and reflect upon the loss of control they sense concerning the future" (Haley 1996: 285). In the native Americas, concern about the end of the world is similarly sometimes linked to a history of colonialism as well as ongoing dispossession and violence: according to Lucas Bessire, for instance, so-called "apocalyptic futurism" among the recently contacted Ayoreo peoples of the Paraguayan Gran Chaco has become "a potent explanatory framework for understanding or creating past events," one that essentially "translates extreme experiences of violence into the conditions for human existence." As Fenn (2003: 110) poignantly put it, "For those who find the present world offering too little in the way of satisfaction or too much in the way of uncertainty, apocalypses satisfy the desire to wipe out the world and, with it, the last vestiges of the singular self."

In short, it seems that apocalyptic futurism among the Urarina might usefully be understood in part as a response to colonialism, violence, and the deep sense of uncertainty that accompanies a rapidly changing world. A sense of disjuncture between the generations, reflected in everything from musical tastes to aspirations to styles of speaking, no doubt exacerbates a certain sense of loss in the face of an unknown future. Older people, in particular, often comment that the end of the world is directly tied to the demise of the powerful shamans of times past; the younger generation is, moreover, simply unwilling or unable to replace them. They are widely perceived to lack, in particular, the physical discipline required to diet or abstain for prolonged periods of time, to build up potency. "They're not strong," I was told. "They'd rather listen to cumbia on their stereos." Yet, while factors such as these contribute to rendering this particular historical moment as one of crisis, a range of other factors must also be taken into account, ranging from people's everyday experience of their immediate physical environment, to the lived experiences of decay and how this informs temporalities, as well as the ways in which the Urarina people assert their own sense of temporal agency.

LIFE IN A FLUID COSMOS

Like many other native Amazonian peoples the Urarina live a mostly subsistence lifestyle, based around hunting, fishing, and small-scale slash-and-burn agriculture, supplemented by occasional work for itinerant traders. Houses are relatively simple structures, built from materials found locally and somewhat open to the elements, as very few have walls. Most travel is by dugout canoe, although these days often powered by small outboard motors. Their territory is remote, far from any towns or urban centers, and the rivers on whose banks they build settlements are dark and sinuous, leading nowhere, meaning that visitors are few, for no one passes through on their way somewhere else. The land is flat and low-lying with abundant swamps; the clayey earth quickly turns to mud in the heavy rain.

This is a fluid cosmos. Bodies and flows of water orient people's lives, beginning with the rivers themselves, on whose banks they dwell and which provide sustenance as well as the primary sense of spatial orientation, in the all-pervasive distinction between "upriver" and "downriver." Cosmogonic myths tell of a primordial flood that marked the beginning of the current era; the existence of such myths are easy to comprehend once having witnessed the torrential rains which can all too easily appear as though they're never going to cease. All this water that passes through Urarina territory both comes from, and eventually ends up in, *cana temura* (literally "our source"), an infinite expanse of water said to encircle the land on all sides and where the terrestrial plane joins the celestial. In short, water is all around, above, and below; people effectively inhabit a tiny patch of land poised precariously above an inconceivably vast ocean, and are continually inundated by yet more water from above. These flows of water are the source of life but they also degrade, corrode, and transform: not least the river itself, whose banks are perpetually washed away and re-deposited, with the effect that the world's primary spatial axis is itself in constant movement.

Given such a fluid cosmos, it is perhaps little wonder that people affirm that when the end of the world comes the sky will fall, the land will sink, the rivers will transform into mud (or fat), and everything will gradually turn into liquid. "All that is solid melts into air," to coin a well-known phrase (Marx and Engels 2010 [1848]), suggesting the power of liquefaction as a metaphor for the erosion of old certainties.

In fact, the world itself is thought to be in a highly fragile and precarious state; not only in a constant state of decay, but perpetually on the brink of collapse. What is about to collapse is, more specifically, *cana cojoanona*, a pivotal concept that literally means "our days" and is the backbone, as it were, of Urarina cosmology.[1] Loosely, *cana cojoanona* is perhaps a little similar to the English term "world," in the sense of "that in which we dwell," or the earthly state of human existence, and indeed is occasionally translated into Spanish by the Urarina themselves as *mundo*. But more often it's translated as *nuestro tiempo* ("our weather" or, alternatively, "our time"). This is felicitous because Spanish, like Urarina (and indeed many other languages), uses the same word for both "time" and "weather,"[2] pointing of course to an intrinsic connection between these two phenomena that is largely lost to those of us living lives largely sheltered from both the fluctuations of the weather and its importance in marking the passing of time. For the Urarina the notion of weather–time clearly has a kind of epochal quality—the "current era," as it were; we might also think of it as akin to what Tim Ingold calls the "weather-world," in which we are immersed and in which nothing stands still, where "every being is destined to combine wind, rain, sunshine, and earth in the continuation of its own existence" (Ingold 2007: S20). But the concept designates more than just a horizon of being, or precondition for life, because humans are also involved in its ongoing production.

The most salient characteristic of *cana cojoanona* ("our epoch" or "our weather") is its extreme fragility. By all accounts it's been in a precarious state at least since the time long ago when the world first "ended." As my friend Bolon once put it, the climate is literally "hanging by a thread," swinging to and fro like a hammock. Here are his words in full:

> Before, in the time of our creation, the world ended. Darkness fell and there was permanent night. It's said there was an old drinker of ayahuasca. It's said that our animals weren't there anymore. The land was completely devoid of animals. Day never broke, it was never dawn. That drinker of ayahuasca, he drank and he saw that the world was ending, without dawn, and suffering with ayahuasca. He was the defender of the world. Because of him, dawn broke. And the animals began to walk around. Those who are seeing the world today—they will see its end. Thanks to that drinker of ayahuasca the world is still hanging by a thread, swinging to and fro. Truly, our lives are tremendously fragile. It's in great danger now, our world.

It seems to me that the fragility of the world and its imminent end is in many ways a logical extension of the phenomenological experience of decay and decline that pervades everyday life. What I mean is, as I've mentioned, everything deteriorates in the fluid Urarina cosmos. In fact, most things in people's lived experience deteriorate remarkably rapidly. The Amazon rainforest is a harsh environment, let's remember: when I once commented to a mestizo trader how quickly the Chinese-made stereos he sold to my Urarina friends seemed to break down, he joked that he could sell them a perfect sphere made of steel and it would break before too long. It's just how things are in this part of the world. And this is one reason, conversely, why highly durable objects such as stones are ascribed extraordinary power and value; or why the most prized types of wood are inevitably those that are hardest and thus slowest to rot. Yet stones, too, like wood and everything else, are eventually corrupted. In this view it seems logical enough that the world, our epoch, would also be subject to degradation. Perhaps the degradation rate of the world is to that of a stone what that of a stone is to that of a fruit. In other words, in short, cosmological entropy would seem to have a phenomenological basis in the pervasive everyday experience of disintegration and decay.

MORAL DECLINE AS CLIMATIC DISINTEGRATION

If everything eventually perishes in this fluid environment—and often sooner rather than later—that is not to say the process of deterioration is entirely outside human control. Steps can sometimes be taken to slow or even momentarily halt the process, from adding salt to meat or fish, to wrapping especially delicate or valuable items, from radios to feather headdresses, carefully in cloth or plastic to protect them. So even entropy on a cosmic scale is not entirely divorced from human action.

Take the example of rain. Even if it is a necessary part of life, people generally dislike rain and speak of it negatively. This is unsurprising if we consider how rain halts the time of everyday life, preventing people from hunting, gardening, or working, forcing them to sit idly in their houses, which no one particularly enjoys. Rain is sometimes referred to as *cana choaje necalano*, literally "filth in the sky" or "filth overhead". When it rains for several days continuously, people start to talk about the end of the world growing near. Strong winds, too, threaten to finish off the world, and thus can frighten people. It becomes harder to find animals in the forest, and there's a general sense of scarcity and deprivation. But rain can also sometimes result from human carelessness. Bathing in the river right after planting manioc, for instance, is said to result in rain. A friend of mine once admitted to me his responsibility for a severe rainy spell, explaining that his wife had inadvertently washed in the river an old rag he had used to blow his nose while drinking ayahuasca the night before.

The gradual decline of game in the forest, of which the Urarina are all too aware, is linked by them less to overhunting, or the introduction of shotguns, than to a general sense that the climate itself is changing and declining over the long term. My friend Lucho would often complain that when he was younger the seasons were regular and one could rely on certain ecological processes to mark out a calendar of sorts: certain flowers and fruits could be expected to appear at the same time each year; and so on. But this was no longer the case, for the seasons are growing increasingly chaotic. This recognition of climate change is closely linked by the Urarina to the growing fragility of the climate itself.

More severe lapses in people's behavior are thought to accelerate the process of disintegration, drawing the end nearer still. One person explained it as follows: "when we eat together, we live like real people. But when our thoughts are different [i.e., when we're stingy], when Our Creator scolds us, the world that he gave us will end." Or, as someone else put it, "Some people go around attacking each other, exterminating each other, thinking differently." But one who sees that all is well in his life, he can say, "no, that's not it." He can say, "that's not for our own good. The world will end in this way, that which Our Creator gave us. If we carry on like that we'll all be lost, together with our world. Everyone."

Closely linked to declining morals and a deteriorating climate is the appearance of demons known as *coitecuchui* (literally "the unknown people"). It seems to me that their presence is somehow linked, in the Urarina imaginary, to the act of incest. In the words of Tivorcio:

> Those evil ones, those demons, the unknown ones (*coitecuchui*), they're going to arrive here. They look just like people. "Where do they come from," we can ask, "these people who we don't know?" And later, the one who knows a little, the one who drinks ayahuasca, can say, "those ones who we don't know, they're *taebuinae* ('savages, demons, bad ones'). They're coming here where we live, it seems they're coming to talk to us, but they're coming to kill us, to eliminate us." That's what we all can say. And so it was for the people who know a bit, the old people, the people who were before us, and so it will be for those who follow us, those who replace us on earth. That's what the ancients said, speaking of all the demons, all the ghosts, all the unknown ones. They are already arriving in some places. This is what people are saying.

Discussing this later with my friend Elias, who helped me translate this passage from the original Urarina, he added some thoughts of his own:

> For example, they've been arriving in [the community of] Pijuayal, Dos de Mayo, and talking to the people there. "I'm your friend, I'm just like you, I'm also like this." That's what they say. The demons arrive where the people are having sex with their mothers-in-law, where they're having sex with their mothers. That's where they're arriving, and that's where the people are talking about it.

Elias went on to say that when he was in the community of Nueva Esperanza a while back the people there warned him that some demons had recently arrived in the community of Dos de Mayo, and had wanted to give out tins of food among other things, before simply disappearing right in the middle of the football field. Others disappeared into the ground and vanished. He was told that the frogs there were speaking, just like people, calling out, saying, "I'm your friend, I also have my family!" and suchlike. "When the demons talk like that, it always rains, every day," Elias told me. "The weather is terrible."

Speaking again of the demons that will roam the earth, one man told me, "When the world is going to end, just a few days before, the *coitecuchui* will come out and burn the world. And they'll say to us, 'you have left God, and you want to have sex with your family'." Some people are thus understandably on the lookout for such demons in everyday life, along with other strange and disturbing occurrences that signal to people that the end is near. One such sign is that animals start talking—comprehensibly, that is, in the Urarina language. Thus one man told me of his alarm when he had a dream in which two cows started chatting to him. There is, it might be pointed out, a throwback here to mythical times, to a now distant past when animals communicated readily with humans and were not yet fully differentiated.

Ideas such as these can easily come to color everyday experience. To give another example: in June 2017 the village was abuzz with rumors and a general state of anxiety after my neighbor Castañon, a quiet and serious man in his late forties or early fifties, was returning home in his canoe late one night when he saw two people swimming across the river. Notwithstanding the fact he was returning after an evening spent drinking manioc beer, the occurrence was soon connected to another rumor then doing the rounds, according to which a group of people from the upstream community of Pijuayal had been drinking manioc beer when two young men they didn't recognise, apparently Urarina, appeared in their midst, seemingly from out of nowhere. When someone from the group asked where they were from, they replied that they were from the community of Santa Rosa de Siamba, which is about a day's travel by boat upstream. The group invited the men to come drink with them, upon which they suddenly ran away into the forest—something everyone found most unusual. The next morning the group called Santa Rosa de Siamba by two-way radio to ask about the men, but their interlocutors knew nothing and said that all members of the community were there at present, that no one was traveling or missing. Different people drew different conclusions from this strange event, but at least one person I spoke to concluded that these two strange men were probably demons heralding the imminent end of the world.

We return here, of course, to the millenarian and potentially Christian elements of the apocalyptic vision that I mentioned earlier. Here is how one person conveyed to me their fears:

If the earthquake doesn't punish us, the lightning will, burning the land. This punishment will affect the "sinners" who have sex with their family. Mother, aunt, sister. God will choose the "good" men and take them to the sky. The sinners will sink with the land. The canoes will transform into alligators, which will attack the men, as will the demons (*coitecuchui*), and the mothers of the forest, of the great trees. The demons will transform into the shape of people and come to converse with those who commit incest.

Consider, too, the following story, told to me by Martin the day after he drank ayahuasca, when I asked him what he saw in his visions. He replied that he was advised that God wishes to "collect" all the people who drink ayahuasca and who believe in him; that he "doesn't agree" with people staying here in the world anymore. "It's not like it was before," according to God. The world is approaching its end, and when all those who believe in him have been removed to safety, to be with God, the land will sink and with it all "those who believe in the inferno." For what it's worth I suspect this is a syncretic vision, mixing new and old beliefs: that in the absence of any mass conversion to Christianity, such a discourse would only spread if it resonated with earlier ideas and precepts. In any case, many people told me that in one way or another poor behavior will bring about the final end. In short, the decline of the climate is insepara- ble from a general condition of moral decline. At the same time though, this helps to shift the burden of responsibility to humans themselves. Like good behavior, in other words, a good and stable climate is poten- tially within human control.

The constant threat of declining moral standards appears consistent with a broader sense of continuous loss that extends well beyond the physical landscape. As Ballard observed for the Papuan Huli, "Entropy might appear somewhat glib as a label for the wide range of attitudes, perceptions, and dicta that are articulated by the Huli on the general topic of decline" (Ballard 2000: 207). Prominent in many Urarina people's minds, as noted above, is the dissipation of knowledge and wisdom resulting from the demise of the great drinkers of ayahuasca of the past. Young people seem too weak and lacking in self-disci- pline to ever replace them, and great uncertainty surrounds the cur- rent moment of rapid social and cultural change. People are scornful of city life in particular: when compelled to travel to urban centers, for instance, to sell forest produce or embark on bureaucratic tasks, they frequently complain about the lack of sharing, or the shameless stinginess of local residents: how no one ever invites them to drink plantain soup or manioc beer; how one cannot do anything if one lacks money. With mounting anxieties about an uncertain future in which even the knowledge of the ancients seems an inadequate guide, it's difficult to escape the sense that people are headed for a time of darkness in which the light of knowledge grows ever dimmer. And yet, there is still hope.

Repair and Redemption

The Urarina people often say that a giant jaguar lives in the sky, and that it regularly fights with the moon. This is the usual explanation of the moon's blood red color at times of lunar eclipse. As the jaguar scratches and claws it the blood flows forth, resulting in a dangerous time for all, for if the jaguar kills the moon it will fall from the sky. The jaguar will follow it down to earth, and when it encounters humans it will easily finish them off. Similar beliefs appear to have been held by many other Amerindian peoples, including the Mayans and the Incas, who would apparently take up their spears and shout at the moon to keep it away (Lee 2014). The Urarina people also saw scope for human intervention: I was told that at times of lunar eclipse the ancients would all go outside in an attempt to save the world from destruction, praying that the moon not fall to earth. Some would shout and beg, others would shoot blowpipe darts, specially "blessed" by specialists, at the jaguar to frighten it. Sometimes, I was told, they would enter its flesh, scaring it away. My friend Martín told me that his grandfather would "bless" ashes from the fire, placed in a bowl and covered carefully with cloth, and then send the womenfolk out to throw it at the sky, until all the ashes were finished. In this way, he said, the sky "normalized" once again.

Ultimately, however, many agree that only God can choose to postpone the end of world, and persuading God is best done by shamans in their state of trance. Thus I was told that "when the ones who drink psychotropics no longer exist, there will be no food to eat, nothing at all, people's hearths won't even light, and so everyone will die." Even those who live today are but a pale shadow of the ancients. According to old Custodio, many are charlatans, who fail to diet properly, or who have sex with mestizos. Only when the blood is pure can psychotropics have their full effect; otherwise, before you know it the lightning will rain down and scorch the earth, until even the rivers will burn, and we will all be lost. Hence the need for people to live well, that is, morally: to share their food, to invite their neighbors to eat with them, to get along with others.

Catastrophic climate change is ultimately less a technical than a social and moral problem. As Ballard (2000: 206) writes, one of the distinctive features of any eschatology is that it provides the universe not just with a sense of trajectory, but with particular prescriptions for moral engagement with it. The idea of the apocalypse unloads a moral burden and imposes it with urgency.[3] Drinkers of ayahuasca, in particular, can still act as a stabilizing force, counteracting cosmic and moral entropy, much as they did when the world almost ended so long ago:

> Before, at the time when the world-era was almost finished, when the world was trembling, those good thunder-people were here, and they repaired it, they saved it. It's because they repaired it that it's lasting up until today. But it's almost about to end, just like that. That's why, because it wants to end, the drinkers of ayahuasca continue maintaining it, maintaining it just there, and because of that, thanks to them, the world continues.

Urarina shamanism is largely a set of techniques for the promotion of health, fertility, and world renewal, in which the tide of cosmic decline is momentarily thwarted. Drinkers of ayahuasca are thus sometimes referred to as *cana cojoanona nunera* (literally "our world-era's support"), and are widely seen as moral beacons on a cosmic scale, courageous as well as wise and persistent, leading a highly disciplined lifestyle in the pursuit of the preservation of life. In the altered states of consciousness induced following the consumption of ayahuasca or brugmansia, they are uniquely capable of requesting Our Creator to bring about replenishment of animals in the forest, the health of ill people in their care, fine weather, and the postponement of the apocalypse. That they ask for all these things simultaneously indicates the extent to which they are interconnected in Urarina thought, perhaps inseparable. Well-being, natural abundance, time, and the weather are all forms of the common good, not only held in common but continuously and collectively produced.

CONCLUSION

In the Urarina language, as in many of the world's languages, the same term can refer both to "weather" and to "time." In most European languages today where this is the case the terms are effectively homonyms—that is, the same word can mean either, but not both at once, even though these meanings were once presumably inseparable. This makes sense, given the experiential connection here between the passing of time as marked in the change from day to night and from season to season. In fact, it seems plausible to speculate that the gradual semantic split between "time" and "weather" may have much to do with the invention of modern time keeping: that with the rise of the clock in the early modern period time became something abstract and absolute, disconnected from the cycles of nature (Smith 2011). The Urarina remind us that, ultimately, weather *is* time, or rather time is a form of weather. In this sense, acknowledging some scope for human intervention in the weather—something altogether common throughout human history—is not entirely different from acknowledging the possibility of intervening in the passing of time itself. In the wake of recent attempts to move beyond the nature/culture dichotomy so long taken for granted in Western thought, I do wonder then about the possibilities for taking seriously a notion of anthropogenic time, or the ways in which time is manipulated or otherwise brought within the remit of human agency—what Ringel and Morosanu (2016: 17) have recently termed "time-tricking," or "the many different ways in which people individually and collectively attempt to modify, manage, bend, distort, speed up, slow down or structure the times they are living in." The kind of time-tricking we are dealing with here seems an ancient one indeed. Michel Serres (1995: 48) writes:

> In the temples of Egypt, Greece, or Palestine, our ancestors, I believe, used to sustain time, as if they were anxious about possible gaps. Here we are today, worried about disasters in the aerial protective fabric that guarantees not time passing, but the weather. They used to connect, assemble, gather, lift up, never ceasing all day long, like monks. And what if it turned out that human history and tradition exist simply because men devoted to the longest term conceivable have never stopped sewing time back together?

We return here to the challenge presented by climate change, as causing what Chakrabarty (2009) describes as "the collapse of the age-old humanist distinction between natural history and human history." An intervention into the climate is an intervention into time itself, and vice versa.

Yet, for all the possibilities afforded by human temporal agency, everything deteriorates sooner or later in the fluid Urarina cosmos. This is a harsh environment, remember: the general pattern is simply for things to decay and degrade sooner or later. In this view the weather-world, our epoch, as a kind of massively distributed hyperobject, in the sense proposed by Timothy Morton (2013), can also be said to be subject to degradation. Cosmological entropy, in other words, has a likely foundation in the routine phenomenological experience of disintegration and decay, in an environment in which nothing stays still.

Let me finish by returning to the concept of *cana cojoanona* ("our world" or "our epoch"). I have described how it embodies a sense of both time and weather, but have said little about the first-person plural possessive marker. What precisely makes this world or epoch "ours," that is, proper to humankind? I can only speculate, but let us consider again the notion that everything decays at its own rate including the world itself. We might say that every object has its own eschatology, its own temporal horizon. This is clear enough in the case of *cana cojoanona*—the world literally *is* time, and time is always directional, always a measure of entropy. But the world decays just like everything else, albeit at a different rate. We might say that every entity has its own time, both in a physical and in a deep ontological sense, and therefore its own eschatology. Put differently, time is perhaps deeply relative, and not a neutral container through which persons or objects drift, so much as an emission of objects themselves (Morton 2013). In the absence of modern time-keeping devices that might give rise to a sense of abstract time, there is nothing outside of the life span of each concrete entity. The immense timescale of the world itself is not that of human affairs; climate is the life cycle of the cosmos. Outside the world, differentiating between now and then is meaningless; an event cannot straightforwardly be specified as happening at a certain place or at a certain time. This is, perhaps, why people can say that the world has in a sense already ended. A single cataclysmic event so far as the world is concerned is massively distributed through time from a human perspective. The shadow of the apocalypse looms out of the past as well as the future, in a kind of temporal foreshortening, with its wake of causality flowing into the present from both directions, threatening but never entirely overcoming the possibility of courageous action by humankind to save it.

Acknowledgement This research has received funding from the European Research Council under the European Union's Horizon 2020 research and innovation programme (grant agreement No 715725).

NOTES

1. The associative prefix "*co-*" links *janonaa* to the first person plural pronoun *cana*, with resultant vowel spreading across the "*j.*"
2. Ingold (2012: 76) points out that several semantically distinct words have a common root in the Latin *tempus*, including the verb *temperare* ("to mix") (from which both *temperature* and *temperate* derive), as well as *temper* and *temperament*, words used to describe human moods and dispositions. "The blending of these different roots is indeed no accident," he writes, "for the weather is a phenomenon of both time and mixture, and of both our affective lives and the aerial medium in which these lives are led."
3. Amanat (2002: 4) argues that an apocalyptic calendric may be "far more essential for the continuity of the sacred and its perpetual renewal than our modern utilitarian notion of linear time and the concept of progress. Time cycles thus may be seen as regulatory means of placing utopian and eschatological aspirations, and whatever is associated with the Beginning and the End, within a humanly conceivable time-frame."

References

Amanat, Abbas. 2002. "Introduction: Apocalyptic Anxieties and Millenial Hopes in the Salvation Religions of the Middle East." In *Imagining the End: Visions of Apocalypse from the Ancient Middle East to Modern America*, edited by Abbas Amanat and Magnus T. Bernhardsson, 1–22. London and New York: I.B. Tauris.

Ballard, Chris. 2000. "The Fire Next Time: The Conversion of the Huli Apocalypse." *Ethnohistory* 47 (1): 205–225.

Bessire, Lucas. 2011. "Apocalyptic Futures: The Violent Transformation of Moral Human Life Among Ayoreo-Speaking People of the Paraguayan Gran Chaco." *American Ethnologist* 38 (4): 743–757.

Chakrabarty, Dipesh. 2009. "The Climate of History: Four Theses." *Critical Inquiry* 35: 197–222.

Danowski, Déborah, and Eduardo Viveiros de Castro. 2017. *The Ends of the World*. Cambridge: Polity Press.

Fenn, Richard. 2003. "Apocalypse and the End of Time." *Daedalus* 132 (2): 108–112.

Haley, Nicole. 1996. "Revisioning the Past, Remembering the Future: Duna Accounts of the World's End." *Oceania* 66: 278–285.

Harkin, Michael E. 2012. "Anthropology at the End of the World." *Reviews in Anthropology* 41: 96–108.

Ingold, Tim. 2007. "Earth, Sky, Wind, and Weather." *Journal of the Royal Anthropological Institute* 13: S19–S38.

Ingold, Tim. 2012. "The Atmosphere." *Chiasmi International* 14: 75–87.

Lee, Jane. 2014. "Lunar Eclipse Myths from Around the World." *National Geographic*. Published 14 April 2004. Available at http://news.nationalgeographic.com/news/2014/04/140413-total-lunar-eclipse-myths-space-culture-science/.

Lewis, Simon, and Mark Maslin. 2015. "Defining the Anthropocene." *Nature* 519 (12 March): 171–180.

Marx, Karl, and Frederick Engels. 2010 [1848]. *Manifesto of the Communist Party*. Marxists Internet Archive.

Morton, Timothy. 2013. *Hyperobjects: Philosophy and Ecology After the End of the World*. Minneapolis: University of Minnesota Press.

Neves Marques, Pedro. 2015. "Look Above, The Sky Is Falling: Humanity Before and After the End of the World." *e-flux* (online journal). Available at http://supercommunity.e-flux.com/texts/look-above-the-sky-is-falling-humanity-before-and-after-the-end-of-the-world/. Accessed 15 September 2017.

Ringel, Felix, and Roxana Morosanu. 2016. "Time-Tricking: A General Introduction." *The Cambridge Journal of Anthropology* 34 (1): 17–21.

Serres, Michel. 1995. *The Natural Contract*. Ann Arbor, MI: University of Michigan Press.

Smith, Justin. 2011. "Being and Weather." *The New York Times*, 29 August.

The End of Days: Climate Change, Mythistory, and Cosmological Notions of Regeneration

Stefan Permanto

INTRODUCTION

Throughout history humankind have been fortunate enough to survive a great number of allegedly apocalyptic events predicted by doomsday prophets from all over the world. I am sure we all remember the times prior to December 21, 2012 when the Maya people were caught in the limelight of a rather dubious end-of-the-world prophecy. As we can conclude, however, this so-called Maya prophecy, heralded almost exclusively by non-Mayans (see, e.g., Restall and Solari 2011), did not bring about the end of the world or any other conspicuous transformation of the planet. The earth stills roams through space about its orbit and we are still here alive and kicking—humans along with all sorts of non-human beings. Still, we may all agree that the Earth has seen better days and that we tend to kick at it rather too hard. Indeed, there are many of us who hold a rather grim view of the future to come. While nuclear

S. Permanto (✉)
School of Global Studies, University of Gothenburg, Gothenburg, Sweden
e-mail: stefan.permanto@globalstudies.gu.se

© The Author(s) 2019   71
R. Bold (ed.), *Indigenous Perceptions of the End of the World*,
Palgrave Studies in Anthropology of Sustainability,
https://doi.org/10.1007/978-3-030-13860-8_4

war has long been a major threat to the world and humanity, today as some have suggested we have entered the Anthropocene and the number one threat to human existence and the environment is climate change. As Samuel Weber (2015: ix) notes, "the apocalypse is in fashion."

Apocalyptic ideas associated with climate change are present among the contemporary Q'eqchi' Maya people in Guatemala and Belize. As one Q'eqchi' elder by the name of Don Félix said: "We are awaiting our deaths; this is the end of days." The aim of this chapter is to scrutinize what (on earth) may have triggered such a dystopian premonition. One thing for certain is that, even though he uttered these words prior to 2012, his statement has nothing to do with the so-called Mayan prophecy. Neither is it linked to any other specific calendar end date inscribed on some ancient stela. On the contrary, as I argue, this eschatological proclamation is implicitly grounded in Maya mythistory and cosmological notions of regeneration and present-day experiences of changes in weather and the environment.

The data I rely on in this chapter stem mainly from my work with an organization called *Li Cheq'il Poyanam* ("The Elderly People") constituted by close to 100 elderly Q'eqchi' men and women, from some 16 different communities scattered throughout the northern parts of Alta Verapaz in Guatemala. It also stems from fieldwork conducted among the Q'eqchi' elders in southern Belize. When I first came to work with the Ch'eqil Poyanam in 2005 they had recently come together as a collective to increase their chances in confronting the local municipality with specific demands and improving their material conditions. As a result of them coming together and working as a unit they have managed to obtain support in acquiring water tanks and *laminas* ("corrugated tin sheets") for roofs, and some of them are involved in projects concerning legal titles to land. Despite these efforts, however, many of the elders feel they have lost the authority and respect that they enjoyed in the past. They say that they feel ignored and that people do not listen to them anymore. They are above all afraid that their cosmological knowledge and *costumbres* ("customs"), passed down to them from their ancestors, run the risk of dying with them. This fear is not unique to the Cheq'il Poyanam and has been observed among other groups of Q'eqchi' people (Carter 1969; Hernando Gonzalo 1997) as well as in other parts of Guatemala (Montejo 2005: 139). As Nygren (1998) observes, the need for indigenous people in Latin America to retell, or re-create, a cultural past constitutes an important part of post-colonial/post-war struggles for ethnic identity, social representation, and self-determination. As I will show here, this is also important for constructing narratives and strategies to cope with climate change.

As several Q'eqchi' elders have told me, they are the bearers of inherited ancestral ritual and cosmological knowledge—knowledge and practices that are crucial for maintaining cosmic equilibrium, peace, and *tuqtuukilal* ("harmony"). Recently, many of them complain that the younger generations are losing interest in the "old ways," which threatens this cosmic equilibrium. They claim that they already see devastating signs of this, particularly in changes to the weather and environment. To counteract such developments the Q'eqchi' elders aim to share their knowledge with younger generations, hoping that this might restore cosmic equilibrium and avoid an eventual Armageddon.

Since culture is neither static nor homogeneous the work and ambition of the elders to preserve and transmit their ancestral knowledge is simultaneously a process of re-discovering and re-creating their cosmological roots and ways of being-in-the-world. Consisting of elderly men and women from several Q'eqchi' communities in Guatemala and Belize, it would moreover be hard to conceive of them as steeped in one and the same "tradition" regarding anything from cosmological notions to ritual practices. The truth is that, even if they share much in terms of culture and history, many of the elders differ in opinion on a range of diverse matters. The Q'eqchi' elders constitute a rather heterogeneous assemblage that to some extent expresses disparate cosmologies, religious beliefs, and enactments of ritual practices, considered to have been inherited from their ancestors. As these are my focus, it is those who belong to the Roman Catholic Church and who generally adhere to what they call a traditional Mayan faith, who take up most space throughout the text. Thus whenever I speak of alleged traditional beliefs and practices this information stems first and foremost from the Catholic elders. Still, while the Q'eqchi' elders in certain instances may well differ in opinion the *Cheq'il Poyanam* in Guatemala claim that they seek to challenge these differences and establish a common ground to preserve and reinvigorate a traditional understanding of their culture and cosmology. By coming together as one force to talk about and promote their ancestral heritage they feel empowered and want to share their knowledge with the younger generations. In fact, it is crucial that they do so because, as many of them state, we are living in the end times.

THE MYTHISTORY OF THE FIVE CREATIONS

To understand and analyze the fears expressed by Don Félix and his fellow elders it is necessary to delve into Maya cosmology as well as to explore mythistorical trajectories that go all the way back to the creation of the Earth. In addressing this Maya cosmogony I draw here mainly on the *Popol Vuh* ("The Book of the People") (Tedlock 1996; Christenson 2007; Goetz and Morley 2003; Edmonson 1971), a Mayan document which was written by the K'iche' Maya people in the early sixteenth century using Roman letters, and is based on an ancient and most likely non-K'iche' pre-colonial codex that escaped the Spanish missionaries' biblioclastic bonfires (even though it is lost today). This document can, moreover, perhaps best be described as a mythistorical pan-Maya document (Tedlock 1996), not only because it describes the creation of the Maya and other neighboring Mesoamerican peoples but also as the stories it contains can be found in myths told by several other groups of Maya people. The Mayan people do not necessarily make a distinction between history and myths, and consequently both may therefore be well grounded in reality (McGee 1989). Moreover, as several Mayanists have shown (cf. Bricker 1977; Burns 1983), historical data can be accurately preserved in oral traditions, not least if they can be complemented with other sources of evidence. As suggested by Tedlock (1996) mythistories consist therefore of both historical and mythical accounts integrated into one balanced whole and as such it conforms to a Maya cosmology that is still valid among many contemporary Mayan peoples. I continue here by presenting in brief how the Popol Vuh describes the creation of the Earth.

As the story goes, at the beginning of time all that existed was a calm sea that lay under a hovering dark sky. The face of the earth (*wach ulew*) was not yet visible. Underneath the surface of the primordial water there existed a being that I refer here to as "Heart Earth" and in the dark sky there resided a deity named "Heart Sky." It is common for Maya deities to have multiple aspects, and they can transform into diverse plural manifestations simultaneously, capable of interacting with one another. As John Monaghan (2000) indicates, Mesoamerican deities are pan-theistic in the sense that they are emanations of a common life force. Such is the case with Heart Earth, which encompasses several paired male and female personifications that communicate and interact,[1] and similarly Heart Sky encompasses three deities.[2]

One day, Heart Sky descended to meet and have a talk with Heart Earth concerning the coming creation of those who would become their future providers and sustainers. Together they pondered on how to conceive light and life and decided the Earth must first be created and sown with life before the dawn would appear. So they set their plan into action. Merely by their words the Earth was created and then *they asked it* to rise from the water, and through their spirit essence (*nawal*) and their miraculous power (*pus*) straightaway mountains (*juyub'*), valleys (*taq'aj*), and forests were *conceived* and came into being which also functioned to set apart the Sky from the Earth.

Having created the Earth and the Sky the stage was set for the Earth to be populated by beings that would venerate and sustain the creators, thus all the animals were conceived. However, as the creators soon noticed, the animals were unable to speak and properly venerate their creators. Acknowledging this mistake they decided to try again to create a being that could properly venerate and remember them, and thus become their provider and sustainer. They created a prototype human whose flesh and body was made out of earth and mud. Regretfully, this being could not walk, multiply, or speak properly and it kept dissolving and falling apart. Heart Earth and Heart Sky undid this creation and tried again. This time they made beings of wood, who looked and spoke like humans and managed to reproduce bearing daughters and sons, eventually populating the whole face of the earth. Still, however, the wooden beings did not possess a heart or a mind, nor had they any blood flowing inside them. They roamed the earth aimlessly crawling on their hands and knees and were unable to remember their creators. Heart Sky decided to let them perish in a great flood. They were also attacked and killed by their pets and kitchen utensils, which complained that for far too long they had been mistreated and disrespected. Some of the wooden people managed to escape and survive, however, and the monkeys that live in the forest today are their descendants. The Popol Vuh mentions the presence of wooden people in the Underworld where they venerated the lords of Xib'alb'a, who shared the same disastrous fate as their subjects.

After this succession of failures the creators tried again. With the assistance of four animals yellow and white corn was discovered inside a mountain. This constituted the flesh or body of the new human

and water was used to make its blood. As when the Earth was created, humanity was endowed with "the miraculous power and the
spirit essence" (*pus* and *nawal*) of the creators and they "became"
alive because of the creators' breath, or interior spiritual force. This
successful fifth and *current* creation resulted in the *hombres de maiz*,
modern humans, who would provide sustenance to the divine beings.
Subsequent to this creation the world saw the rising of the sun lighting
up the world.

THE TZUULTAQ'A

I will now recapitulate by traveling farther back in Maya mythistory to
the moment of the creation of the earth, or as it is called in the Popol
Vuh, *Juyub' Taq'aj* (literally "mountain/hill valley"). Clearly, this earth
was considered an animate being, infused with an animating force,
since (as mentioned earlier) it was *asked* to rise from underneath the
surface of the primordial water, which indicates that it in some sense
was able to hear and act accordingly. It is, moreover, common knowledge among many contemporary Maya people that the earth is a living being. The Q'eqchi' people refer to this living earth as *Tzuultaq'a*
("mountain/hill valley), which in line with the common Mayan cosmological notion of complementary opposites also means "the one and
the many." While manifesting as *one* singular being—the living earth
or "Mother Earth" (Sp. *madre tierra*)—the Tzuultaq'a manifests simultaneously as many individual beings (referred to here as tzuultaq'as)
linked to specific hills, mountains, and caves. In this plural form,
tzuultaq'as are equivalent to the Chacs, Bacabs, and Pahuatun, ancient
Maya weather-deities associated with caves and mountains (Thomson
1990). Such "earth beings" or "lords of the hills" are still of great
importance in the lives of many Maya people throughout Guatemala,
Mexico, and Belize. According to Q'eqchi' myths these tzuultaq'as
walked the Earth prior to the fifth creation when the sun had yet to
ascend to the heaven and light up the world. While these beings are
equivalent to ancient weather deities, I suggest also that they correspond to the creators of humankind.

For many contemporary Q'eqchi' people the tzuultaq'as are considered to be God's sentinels stationed here on Earth. Tzuultaq'as are

sometimes metaphorically likened to a foreman at a rural *finca* while (the Christian) God represents the owner who lives far away in the capital. Perhaps we could say (and I think we can) that God is like a distant Heart Sky, resident in the Heavens, while Tzuultaq'a constitutes the terrestrial Heart Earth. Tzuultaq'as roamed the Earth prior to the creation of humanity. Similar to the stories in the Popol Vuh, Q'eqchi' mythistories tell of how the tzuultaq'as managed, with the assistance of animals, to find the hidden corn within a mountain. Grandfather Xpiyakok and Grandmother Xmukane (see Note 1) are two manifestations of the creators mentioned in the Popol Vuh and they are equivalent to Itzamna and Ix Chel/Chak Chel, who I argue were earth deities (or tzuultaq'as) venerated by the pre-colonial Maya, including the Q'eqchi' Maya (cf. Taube 1992: 92–99; Christenson 2007: 63; Thompson 1990: 242; van Akkeren 2000: 102, 104–105, 235; Tedlock 1996: 217; Bassie-Sweet 2008: 127). I suggest that the tzuultaq'as are (or at least in earlier days were understood by Q'eqchi' people as) the creators of humankind, and the Maya creators were in turn earth lords.

The tzuultaqa'as share the same spiritual interiority as humans, understood to be subjects or persons endowed with the ability to think, reflect, and communicate and to act intentionally (cf. Descola 2013). Kinship, gender, and hierarchy structure their social organization: while they are parents, siblings, and married couples, they are also organized under 13 major tzuultaq'as, of which a "king" and "queen" are the most powerful. As individual beings they exhibit distinct dispositions: some are more easily angered and offended than others and act accordingly, and sometimes they make simple mistakes as we humans do. They live in houses, or stone houses (*ochoch pek*), which refer mainly to caves inside hills and mountains, often described as great mansions. Indeed, every hill or mountain that has a cave inside it is intimately linked to a particular resident tzuultaq'a who is the owner of his or her respective hill and adjacent valleys, which turns the entire landscape into a mosaic of properties owned and controlled by different tzuultaq'as. The tzuultaq'as are considered masters of everything that exists upon and within the earth, which includes everything from corn, beans, bushes, trees, bodies of water, animals, and in a certain sense, human beings.

According to Q'eqchi' myths, in the beginning of time, long before humans entered the scene, corn was not available for consumption. At this early time in Maya mythistory there was a great famine in the world with the result that both animals and tzuultaq'as suffered immensely and were on the brink of dying of starvation (Permanto 2015). Eventually, however, as mentioned above, 13 (or 12) tzuultaq'as eventually found the precious corn hidden away inside a mountain and learned the proper rituals required for enjoying bountiful harvests (Preuss 1993; Burkitt 1920). Having retrieved the corn the king Tzuultaq'a scattered corn seed all over the woodlands so that it could be consumed by animals, and in temporal extension also by the human beings who were yet to be created.

Other myths tell of how the tzuultaq'as became masters of all wild animals. In distant times a hunter and a fisherman called Lord Red Star (*Qaawa Kaqchahim*) was the original owner of wild animals but when he chose to ascend to the sky as the Morning Star (Venus) he left all his animals in the custody of the tzuultaq'as (Schackt 1986: 59–69, 176–179). Thus, ever since, tzuultaq'as keep and care for all wild animals in pens inside their respective hills and mountains and let them out in the forests to be hunted by humans.

Besides being keepers of animals, tzuultaq'as also supervise human moral and ritual behavior. In this sense they care for humans much as Q'eqchi' parents would care for their children. While tzuultaq'as, as individual beings, are often referred to as anything from ancient men and women to brothers and sisters to the Q'eqchi', in its singular manifestation the Tzuultaq'a is conceived of as the mother of the Q'eqchi'. The Q'eqchi' people accordingly refer to themselves as "the children of the [Mother] Earth" (*aj ralch'och*), and in line with the pan-Mayan cosmogony of the Popol Vuh their myths tell of how the Q'eqchi' ancestors originated from a cave within the earth's interior, which is equated with the womb of Mother Earth (Heyden 2005: 22). The Q'eqchi' elders say that they have a strong attachment to their ancient mother earth and that she raises them as her children. They are dependent on her for their survival and sustenance (Permanto 2015: 57), and myths tell of how human infants used to be breastfed by a Tzuultaq'a (Cabarrús 1998). Thus the Tzuultaq'a acts as the protecting mother that she is—nurturing and caring for her children, whilst demanding that people are respectful not only to the tzuultaq'as but toward all beings, fellow humans as well as non-humans.

Just as Tzuultaq'a is the overseer of crops, animals, plants, forests, and human beings, it is also responsible for meteorological and atmospheric phenomena. As the Q'eqchi' elder Don Vicente told me: "The Tzuultaq'a controls rain, clouds, thunder and lightning, hail, and wind." Don Francisco, another elder, said: "When it blows really hard, it is the Tzuultaq'a who is responsible." From a hole in the ground by the road close to Francisco's village the local Tzuultaq'a spends all day giving the air and wind that sustains plants and people. Here the Tzuultaq'a is reminiscent of the ancient Maya rain gods, the Chacs, who in pre-colonial times *personified* not only rain (Thompson 1990) but also wind and hurricanes (Tozzer 1978: 137–138). Prior to planting corn, people frequently ask the local tzuultaq'as for favorable weather conditions, such as that the right amount of rain to come at the appropriate time. Lightning and thunder are also the doings of the tzuultaq'as, resulting from them hurling their axes through the air (Sapper 2000). What we in other parts of the world would understand to comets or falling stars are, according to the Q'eqchi' people, *correos*, or messages, traversing the hills and the valleys between the tzuultaq'as. Similarly, as Estrada Monroy (1993: 298) notes, when a tzuultaq'a visits another tzuultaq'a he or she may do so in the form of a fireball.

EXISTENTIAL RECIPROCITY

There is no doubt that the Tzuultaq'a plays a central part in the life-worlds of many Q'eqchi' elders, and that it is imperative to establish and maintain a harmonious relationship between the two parties. To be allowed usufruct to all the tzuultaq'as' resources, they demand that humans (1) lead a morally correct life respecting all life and (2) conduct the proper feeding rituals directed to the tzuultaq'as, and that they (3) ask for what they wish (e.g., bountiful harvests, to not let animals eat their crops, favorable weather conditions, etc.). I would summarize this relationship between humans and tzuultaq'as in terms of reciprocity, or what I have called an existential reciprocity (Permanto 2015). While the Q'eqchi' people are highly dependent upon the Tzuultaq'a for their sustenance and survival, the reverse is also true. The rituals directed toward the Tzuultaq'a can mainly be understood as feeding rituals in which the Tzuultaq'a consumes everything from animal blood to incense, cacao, and flowers. Accordingly, both humans and tzuultaq'as are inherently dependent on one another for nourishment and sustainment—or life. As noted earlier, this existential reciprocity can be traced to the beginning of creation. In fact, if we would trust the mythistorical accounts this reciprocal exchange is essentially the *only* reason that humankind was ever created at all. Thus, while the relation between humankind and the Tzuultaq'a revolves around reciprocal sustenance, in extension it functions also as a kind of ecological regulator—or, as some Q'eqchi' would have it, a regulator of cosmic equilibrium.

According to many contemporary Q'eqchi' people if they fail to lead a morally correct life, fail to ask for permission to use the tzuultaq'as' resources, or to properly conduct the required feeding rituals prior to hunting, cutting down trees, planting corn, etc., they fail to observe the reciprocal bond that was set up at the beginning of Creation, and this will result in devastating consequences and wreak havoc on the world. Such behavior will unleash the rage of the Tzuultaq'as who may snatch away the spirit of the transgressor and withhold animals and corn seeds within their caves so that they are not accessible for people. They may also send wild animals to eat the crops of transgressors.

Thus failure to uphold the reciprocal bond between humanity and Tzuultaq'a is related to ruptures in moral and ritual conducts, which occurs for a number of reasons. Essentially, however, as many Q'eqchi' elders complain, this rupture comes down to the fact that many people are diverting their attention from the ancient traditions, or *costumbres*. They say that young people, in particular, are following instead the "modern traditions of other foreign cultures." Still, they are also aware that many elders too fail to perform the required rituals even though many of them feel an urgent need to revitalize and re-enact these rites.

Reasons given as to why people fail to lead a morally and ritually correct way of life relate primarily to centuries of displacement, poverty, religious fragmentation, and aspects of modernity (Permanto 2015). With regard to experiences of displacement the Q'eqchi' people have for lengthy durations of time in different periods, such as during the nearly 40-year-long Guatemalan civil war extending from the 1960s to the late 1990s, been forced not only to leave their homes but also their local tzuultaq'as and eventually this has led to many people forgetting both their *costumbres* and their tzuultaq'as. If people have forgotten how to behave ritually this means that they have failed to pass this knowledge along to the younger generations as a result. Poverty is also a major factor since feeding rituals directed to the tzuultaq'as require the purchase of the necessary paraphernalia, which can be rather expensive. Thus lack of financial resources may prevent people from performing the proper rituals. The elders perceive, furthermore, that religious division is causing disruption and friction among communities. While the Maya have been Catholic since colonial days, new Protestant Evangelical churches are currently being established in their communities at an increasing rate. It is not uncommon that one village harboring some 90 households should have 6 or more different churches, of which only one is Catholic. Evangelical priests tend to preach that the Tzuultaq'a is no other than the devil personified and that faith in this evil demon, along with the traditional customs and rituals directed to him, should be wiped from the face of the earth. Still, as many elders have told me, no matter if you are Evangelical or Catholic you still pray to the same God. With regard to modernity, many elders complain that many of the young people of today turn their backs on traditional community life and *costumbres* and instead embrace a non-indigenous modern lifestyle. While many elders do not have anything against development and external modern elements per se, they nevertheless consider them to constitute a potential threat to their *costumbres* and cosmological notions. As Don Santiago claims, for example, "if people stop paying attention to the ancestral ways we may face the same destiny as did the ch'olwinq." The ch'olwinq were ancient Mayas belonging to a previous creation but, because they diverted from their intended way of life, they are now invisible living disconnected from the present society.

These estrangements from traditional beliefs and practices have led to a significant decrease in feeding rituals as well as a rupture in social relations among fellow humans in the communities. Thus the reciprocal bond between humanity and the Tzuultaq'a disintegrates and the Q'eqchi' elders fear that eventually, if this negative trend continues, it will disturb cosmic equilibrium and wreak havoc in the world. They see many signs of such a grim future approaching.

CHANGES IN THE CLIMATE,
THE WEATHER, AND THE ENVIRONMENT

When people hunt wild game uncontrollably without asking for permission and fail to conduct the proper associated rituals, this may well lead to overhunting and eventually result in the loss of all game animals. As a sign of this many Q'eqchi' people claim they see fewer animals and birds in the forest today. In a dream one elder found himself inside the cave of a Tzuultaq'a where he was shown empty pens, and the resident Tzuultaq'a complained that this is what happens when people hunt without asking for permission. All of the Tzuultaq'a's animals had been killed without having been paid anything in return. In a similar manner the elders see disturbing signs of climate change as a direct consequence of the lack of belief in the tzuultaq'as.

It must be mentioned also that tzuultaq'as occasionally affect the environment in negative ways of their own. Flooding, for instance, may be a sign of the *fiestas* that the tzuultaq'as are having in their subterranean domains (Sapper 2000: 33). There is also the story of the furious Tzuultaq'a who in the 1970s flooded a river and caused a terrible storm because his beloved had cheated on him. Trees and houses were destroyed as a consequence of his rage.[3] Rumblings and earthquakes may also result from the activities of tzuultaq'as when they move around inside their hills and mountains.

Nevertheless, the elders claim that many cases of negative climate change stem from breaches in the reciprocal relationship between humankind and the tzuultaq'as. They complain that the weather is becoming increasingly unpredictable, which makes it more difficult to decide when one should start planting, seriously affecting the harvests

(cf. Penados and Tzec 2012). The elders say that it is getting hotter
and drier and fear that the landscape will soon become a desert. They
no longer hear the thunder as often as they used to and there is less
wind and rain, and in some places the coolness of winter is prolonged.
They complain that animals are disappearing from their forests at an
increasing rate, not least because of deforestation. To cut down trees
and even entire forests uncontrollably, without the consent of the pertaining tzuultaq'a, leads to devastating consequences. As one elder told
me, a hill with a lot of trees growing on it is a fresh, vital, and powerful tzuultaq'a, while a deforested hill or mountain is a disempowered or
even dead tzuultaq'a. People tell stories of tzuultaq'as that have chosen
to abandon their hills because people no longer respect and feed them. If
people keep cutting down forests it weakens them, or it means they are
not present at all. Thus, just as scientific climatology holds that deforestation releases greenhouse gases which in turn leads to global warming,
the Q'eqchi' people agree that deforestation (which is very extensive in
Guatemala today) affects the climate, although often for quite different
reasons. Since it is ultimately the tzuultaq'as that are in charge of the
weather, if people no longer take an active part in the reciprocal relationship with them they lose the ability to *indirectly* affect the weather.

Should people abandon and eventually forget the ancient rituals,
which is what the elders fear will happen, this will undoubtedly lead to
the end of the world. Not immediately, perhaps, but, as one Q'eqchi'
elder in Belize told me, "in 30 years people will die of starvation." We
might remember in this context the famine of ancient days when all the
animals and the tzuultaq'as suffered from hunger due to a lack of corn.
Unless people again learn how to gain access to corn by conducting the
proper rituals the bond of existential reciprocity will be broken resulting in disturbances to the cosmic equilibrium. This would cause famine
among humans and tzuultaq'as, and have grave impacts on the environment and the climate.

If humans neglect to ask for permission to hunt animals and cut down
trees, these activities can be carried out in a runaway fashion resulting in
overhunting and deforestation. This would not only lead to a decrease in
the fauna but to uncontrollable changes in weather and climate, due to
the desertion of the tzuultaq'as. I was once told of a terrible hurricane
that swept in over two communities. In one community people venerated and fed the local tzuultaq'a, whereas in the other people did not. As
a result the first community was saved from destruction by the hurricane,

while in the other many houses and cornfields were destroyed. According to the Q'eqchi' elders in the first community this was proof that if one maintains good relations with the tzuultaq'as, one is protected from bad weather. If, however, the reciprocal bond between humanity and the Tzuultaq'a is permanently and universally broken this will eventually lead to the end of the world as they know it. Seemingly, as the elders see it, human beings are failing to respect, venerate, and sustain the deities in the proper way, and if this development is not halted we humans risk going the same devastating way as previous creations (i.e., the animals, the mud people, and the wooden people). If this negative trend continues, the current creation will soon have acted out their last scene in this life drama, posing the question of whether there will be a new creation or if the world will retreat to its primordial condition.

COSMOLOGY AND MYTHISTORIES OF RENEWAL

When asked about the future prospects for veneration of the Tzuultaq'a, several of the elders conjured up a grim and sad future. One believes that when all the elders are gone there will be no one left to perform the rituals. When all rituals are forgotten, he says, humanity will experience difficult times: the harvests will yield nothing, the animals will die, and people will starve. Others are less certain of what the future will bring, since as they say they do not yet know how future generations will look upon the Tzuultaq'a. Even if people experience times of extreme misery, they may eventually realize that it is through the correct performance of rituals that humanity will be able to enjoy a good life with bountiful harvests and will take up these practices again. Many elders are determined, however, to set the record straight and re-enact the ancient rituals in their lifetimes, sharing their knowledge with younger generations. In southern Belize, for example, elders from four villages have decided to take it upon themselves to act in favor of *all* communities in the region and gather annually to venerate and ask the Tzuultaq'a for favorable weather for their crops. Such collective feeding rituals and petitions occur annually, primarily prior to planting corn and other foodstuffs, and thus constitute the beginning of the cycle of agricultural fertility and regeneration (Wilson 1995).

Although many of my Q'eqchi' interlocutors fear that humanity will eventually perish due to the failure of the reciprocal bond between humanity and the Tzuultaq'a, I have never heard them talk explicitly about the possibility of a new creation. Nevertheless, if we view this scenario from the perspective of Maya cosmogonies, it would not seem totally unlikely that mythistory could repeat itself.

The Maya creations are to a certain extent equivalent to those of the Aztecs, where it is said that this sun will end in earthquakes and other catastrophes and eventually result in a new creation. Similarly to the Maya, the Aztec people considered themselves to live in the fifth sun, or the latest successful attempt in creating human beings who were able to appease the deities to prevent another obliteration.

Maya cosmologies incorporate apocalyptic ideas, although these are not overtly linked to any specific calendar date and cannot be used to forecast a specific "end date," such as happened prior to 2012. The Popol Vuh describes apocalyptic events just as much as it depicts cyclical events of renewal. Maya cosmology is fundamentally structured around the phenomenon of renewal and cyclical regeneration where life and death are inextricably and continuously interlocked. Carlsen (1997: 50) notes, "Mayans have believed that life arises from death," a historical observation which is to some extent valid today. This logic applies to humans as well as to maize: when an individual maize plant dies it results in numerous seeds that eventually are returned to the earth where it becomes a new living plant. Maya mythology mentions frequently the existence of beings that belonged to previous creations but that continue to exist within mountains and in the Underworld, much like the wooden people. Somehow they have survived previous apocalypses and live hidden away from the current creations that have replaced them (Cook 2000: 140). The Q'eqchi' people conceive of such a people that may well be linked to the ancient wooden people. They are spirit beings called the Ch'olwinq who live hidden away inside mountains and/or deep in the jungles (Permanto 2015). Redfield and Villa Rojas (1962: 328–335) report similar mythistories, some of which they refer to as being "millenialistic," indicating the re-emergence of ancient beings when humanity is done with (or has rid themselves of) this earth. Perhaps we can say that they are situated in a liminal waiting period. There are similar tales of Andean beings, remnants of previous worlds, who survive in caves and burial towers (Burman 2016). I suggest, moreover, that the Q'eqchi' notion of living in the end times is reminiscent of the Andean concept of *pachakuti* ("time upheaval"), which implies that the tumultuous times we find ourselves in today could in the worst case result either in a cosmic catastrophe or in the re-creation of a new era—a cosmic renovation of sorts (Thomson 2011). Although the Q'eqchi' elders do not explicitly argue for the coming of a new being that will take our place, the Mayan cosmology of regeneration and mythistories of renewal indicate that the world will *continue without us* (cf. Danowski and Viveiros de Castro 2017), with either the re-emergence of ancient beings or by the creation of new improved beings.

How then do we humans expect the world to end? As mentioned at the beginning of this chapter, it seems as though the foreboding apocalypse is associated with changes in the climate. We are expected to suffer the consequences of drought, floods, rising sea levels, as well as storms and hurricanes. The Q'eqchi' elders say that they are already seeing signs of this development that eventually will lead to starvation and the end of humanity. If the earth scientists are right in that we have entered the Anthropocene, it follows that an eventual climatic apocalypse will be the result primarily of anthropogenic actions. Seen from the perspective of the Q'eqchi' elders, however, we might say that an end-of-the-world scenario is the result of a *lack* of anthropogenic performances, specifically neglecting to ritually respect the non-human tzuultaq'as. Perhaps then humanity will meet the same fate as previous creations, and failure to sustain the spirits will end in terrible climatic catastrophes, Maya mythistory will repeat itself, and humanity will perish due to terrible storms and floods.

NOTES

1. Framer (*Tza'qol*)/Shaper (*B'itol*), She Who Has Born Children (*Alom*)/ He Who Has Begotten Sons (*K'ajolom*), Junajpu Possum (*Junajpu Wuch'*)/Junajpu Coyote (*Junajpu* Utiw), White Great Peccary (Saqi Nim Aq)/Coati (*Sis*), Its Heart Lake (*U K'u'x Cho*)/Its Heart Sea (*U K'u'x Palo*), S/he of Blue/Green Plate (*Aj Raxa Laq*)/ S/he of Blue/ Green Bowl (*Aj Raxa Sel*), She Who Has Grandchildren (*I'yom*)/ He Who Has Grandchildren (*Mamom*), Xmuqane/Xpiyakok (Tedlock 1996; Christenson 2007).
2. *U'kux kaj*, or *Juracan*, encompasses three beings: Thunderbolt Huracan (*Kaqulja Juraqan*), Youngest Thunderbolt (*Ch'i'pi Kaqulja*), and Sudden Thunderbolt (*axa Kaqulja*) (Tedlock 1996; Christenson 2007).
3. I received this story from Peace Corps volunteer Amy Olen.

REFERENCES

Bassie-Sweet, Karen. 2008. *Maya Sacred Geography and the Creator Deities*. Norman: University of Oklahoma Press.

Bricker, Victoria R. 1977. "The Caste War of Yucatan the History of a Myth and the Myth of History." In *Anthropology and History in Yucatan*, edited by Grant D. Jones, 251–258. Austin: University of Texas Press.

Burkitt, Robert. 1920. *The Hills and the Corn*. Philadelphia: University Museum.

Burman, Anders. 2016. *Decolonization in the Bolivian Andes: Ritual Practice and Activism*. Lanham, MD: Rowman & Littlefield, Lexington Books.

Burns, Allan F. 1983. *An Epoch of Miracles. Oral Literature of the Yucatec Maya*. Austin: University of Texas Press.

Cabarrús, Carlos. 1998. *La Cosmovisión Q'eqchi' en Proceso de Cambio*. Guatemala: Cholsamaj.

Carlsen, Robert S. 1997. *The War for the Heart & Soul of a Highland Maya Town*. Austin: University of Texas Press.

Carter, William E. 1969. *New Lands and Old Traditions: Kekchi Cultivators in the Guatemalan Lowlands*. Gainesville: University of Florida Press.

Christenson, Allen J. 2007. *Popol Vuh: The Sacred Book of the Maya*. Norman: University of Oklahoma Press.

Cook, Garrett W. 2000. *Renewing the Maya World: Expressive Culture in a Highland Town*. Austin: University of Texas Press.

Danowski, Déborah and Eduardo Viveiros de Castro. 2017. *The Ends of the World*. Cambridge: Polity Press.

Descola, Philippe. 2013. *Beyond Nature and Culture*. Chicago and London: University of Chicago Press.

Edmonson, Munro S. 1971. *The Book of Council: The Popol Vuh of the Quiche Maya of Guatemala*. New Orleans: Middle American Research Institute, Tulane University.

Estrada Monroy, Agustin. 1993. *Vida Esotérica Maya-K'ekchi*. Guatemala: Impreso Serviprensa Centroamericana.

Goetz, Delia, and Sylvanus Griswold Morley. 2003. *Popol Vuh: The Book of the Ancient Maya*. Mineola, NY: Dover Publications.

Hernando Gonzalo, Almudena. 1997. "La identidad Q'eqchi'. Percepción de la realidad y autoconciencia de un grupo de agricultores de roza de Guatemala." *Revista Española de Antropología Americana* 27: 199–220.

Heyden, Doris. 2005. "Rites of Passage and Other Ceremonies in Caves." In *In the Maw of the Earth Monster: Mesoamerican Ritual Cave Use*, edited by James E. Brady and Keith M. Prufer, 22–34. Austin: University of Texas Press.

McGee, R. Jon. 1989. The Flood Myth from a Lacandon Maya Perspective. *Latin American Indian Literatures Journal* 5 (1): 68–80.

Monaghan, John D. 2000. "Theology and History in the Study of Mesoamerican Religions." In *Ethnology*, edited by John D. Monaghan, assisted by Barbara W. Edmonson, 24–49. Supplement to the Handbook of Middle American Indians, vol. 6, Victoria Reifler Bricker, general editor. Austin: University of Texas Press.

Montejo, Victor. 2005. *Maya Intellectual Renaissance: Identity, Representation and Leadership.* Austin: University of Texas Press.

Nygren, Anja. 1998. "Struggle over Meanings: Reconstruction of Indigenous Mythology, Cultural Identity, and Social Representation." *Ethnohistory* 45(1): 31–63.

Penados, Filiberto, and Angel Tzec. 2012. *Indigenous Peoples and Climate Change: Impacts of Climate Change on Food Security in Two Q'eqchi' Maya Communities in Southern Belize.* Belize: SATIIM.

Permanto, Stefan. 2015. "The Elders and the Hills: Animism and Cosmological Re-Creation Among the Q'eqchi' Maya in Chisec, Guatemala." PhD diss., Gothenburg: University of Gothenburg.

Preuss, Mary H. 1993. "The Origin of Corn and Preparation for Planting in K'ekchi' and Yucatec Mayan Accounts." *Latin American Indian Literatures Journal* 9 (2): 120–134.

Redfield, Robert, and Alfonso Villa Rojas. 1962. *Chan Kom: A Maya Village.* Chicago and London: University of Chicago Press.

Restall, Matthew, and Amara Solari. 2011. *2012 and the End of the World: The Western Roots of the Maya Apocalypse.* New York: Rowman & Littlefield Publishers.

Sapper, Karl. 2000. "Religious Customs and Beliefs of the Q'eqchi' Indians." In *Early Scholar's Visits to Central America: Reports by Karl Sapper, Walter Lehman, and Franz Termer*, edited by Marilyn Beaudry-Corbett and Ellen T. Hardy, 29–40. Los Angeles: University of California.

Schackt, Jon. 1986. *One God, Two Temples: Schismatic Process in a Kekchi Village.* Oslo: Department of Social Anthropology, University of Oslo.

Taube, Karl. 1992. *The Major Gods of Ancient Yucatan.* Washington, DC: Dumbarton Oaks Research Library and Collection.

Tedlock, Dennis. 1996. *Popol Vuh: The Definitive Edition of the Mayan Book of the Dawn of Life and the Glories of Gods and Kings.* New York: Touchstone, Simon & Schuster.

Thompson, J. Eric S. 1990. *Maya History and Religion.* Norman: University of Oklahoma Press.

Thomson, Bob. 2011. "Pachakuti: Indigenous Perspectives, *buen vivir*, *sumaq kawsay* and Degrowth." *Development* 54 (4): 448–454.

Tozzer, Alfred M. 1978. *Landa's Relación de las cosas de Yucatán.* Millwood, NY: Kraus Reprint Co.

van Akkeren, Ruud. 2000. *Place of the Lord's Daughter: Rab'inal, It's History, It's Dance-Drama*. Leiden: Research School CNWS, School of Asian, African, and Amerindian Studies (CNWS Publications, No. 91).

Weber, Samuel. 2015. "Foreword: One Sun Too Many." In *Apocalypse Cinema: 2012 and Other Ends of the World*, edited by Peter Szendy, ix–xx. New York: Fordham University Press.

Wilson, Richard. 1995. *Maya Resurgence in Guatemala: Q'eqchi' Experiences*. Norman: University of Oklahoma Press.

Contamination, Climate Change, and Cosmopolitical Resonance in Kaata, Bolivia

Rosalyn Bold

Critics of the "ontological turn" claim that in its commitment to evoke a non-naturalist cosmos the movement excludes indigenous engagement with modernity, preferring to reify "indigenous" or non-modern ontologies and emphasizing their incommensurability with "modernity" (Bessire and Bond 2014; Holbraad and Pedersen 2017). I will show here how melding worlding practices together can contribute to understanding and handling the truly cosmopolitical phenomenon of climate change. Relational landscapes can be of utility to a modern understanding of environment and contamination by expanding isolated variables such as waste plastics or changing temperature into a network of interrelating human and non-human actors. The concept of scale and inter-relationality present in my fieldsite, Kaata, renders climate change relevant, locally experienced, and intrinsically related to the actions of humans within their agentive environments. It thus presents "lines of flight"

R. Bold (✉)
Department of Social Anthropology,
University College London, London, UK
e-mail: r.bold@ucl.ac.uk

© The Author(s) 2019
R. Bold (ed.), *Indigenous Perceptions of the End of the World*,
Palgrave Studies in Anthropology of Sustainability,
https://doi.org/10.1007/978-3-030-13860-8_5

(Deleuze and Guattari 1988) from a modern mentality, while as we shall see it does not ultimately provide a solution to climate change.

The Quechua-speaking farmers of ayllu[1] Kaata, highland Apolobamba, are living a transition from an animate landscape to a de-animated Anthropocene, which whole complex of change they refer to when they employ the Spanish term *cambio climático* ("climate change"). Human actions are networked into reciprocal interactions or conversations between humans and non-humans, and climate change occurs within these networks, affecting everything, touching the reciprocal ties constituting the landscape, and often stretching them to their limits. Climate change thus constitutes a total transformation of the landscape of the community, and indeed could be conceived as a shift in the ontological currents composing it.

Reciprocity, as the linguist Bruce Mannheim indicates, is among Quechua speakers the defining characteristic which sets off *runa* ("human beings") from the uncultured "other"—the whites of the towns and cities (Mannheim 1986). The term *ayni* connotes direct symmetrical reciprocity between equals, often connoting working relationships between *runa*, while *mañay* signifies delayed reciprocity such as occurs in prestations to mountain deities, often accompanied by counter-requests. Mannheim argues that *mañay* has "the character of a total social phenomenon" (1986: 268) in the Andes, thus reciprocal relations are the ontological warp and weft of the landscape.

In 1978 Joseph Bastien, in his now classic ethnography *Mountain of the Condor*, wrote that the Kaatans described the mountain they inhabit in terms of a living body, made up of both human and "natural" features: mountain lakes were its eyes; the main square where rituals were once performed and below which the ancestral mummies are buried was its heart. Kaata's fields lay on the stomach or center of this mountain body, where potatoes are grown and exchanged for meat from the highland herders at its "head." Conducting fieldwork between 2010 and 2014 I was told that "we no longer say this" by villagers, yet the mountain as an animate actor surfaces to challenge modernity.

The continuity of Kaata as an entity is threatened in the current moment—as the older people told me, there are now only a few of them remaining in the village "to maintain the traditions," and my younger friends, back to visit from the city, feared that in a generation's time it would be deserted entirely. Migration to the cities and coca fields is very high, and young people are driven by desire for clothes and city foods, meaning there are few available to labor in the fields.

The villages are famous for their healers and diviners, known as Qollahuaya. There is now only one man in Kaata considered Qollahuaya, skilled in the healing arts, who is also the "Lord of the Seasons," the head of ceremonies. Curing is undertaken through divination, relying on the analogical associations of human and mountain bodies. Bastien (1978) records that to read the misfortune of a plaintiff, coca leaves would be thrown on a mat, reflecting where in the plaintiff's body illness or misfortune was located. The human body was then healed by curing the pertinent point on the body of the mountain. This is a landscape of reflection and interrelation across scales; it is fractal.

THE NEIGHBOR AND THE WIND

In the sunshine of the dry season in Kaata I was speaking to a neighbor of the community leader over his garden wall, while around us the mountain terraces dried to golden dust. He took up the theme of climate change with enthusiasm, and explained with concern that the people of "before" in Kaata "would live until they were 120, and there was no sickness. Now we have more illnesses. I've got a sore throat, for example." "Me too," I said. "In those times," he continued, "everything was different; the sun, winds, air were different."

We see here something of the complete cosmic transformation that for Kaatans comprises "climate change." The "before" the neighbor was tracing is recent—a living man, Don Ramon, by all accounts 96 years old, is cited as an exemplar of these "people of before", strong and hardworking. Moreover, Bastien's fieldwork 40 years ago was similarly classified as "before." In those days, the man went on to tell me, people ate the produce of the place—*oca* ("a tuber"), potatoes, wheat, barley, peas, beans, and maize, while now they eat rice, pasta, and city foods, which don't have the same—and here he paused in an effort to find a cosmopolitical translation—vitamins. "The children are different now," he continued. "We would obey our parents... We used to give our children *tostadas* of maize, beans. They eat yoghurt now, and rice, pasta—those things. You can't stop it." I asked him if, with the government's agricultural sovereignty schemes, which might create cash income from agriculture in the village, inducing young people to stay, climate change might be stopped. The man threw up his hands and laughed. "No, you can't stop it! The contamination is borne on the wind, from other places, big factories…"

"Climate change" here describes a network of actors, human and non-human. It involves great powers, like the sun and winds, for this man it was visible also in the actions of the young people, the transformation of their desires toward commodities which come from outside the community, describing change in the entire landscape (Ingold 2000) of the village. Suffusing this living landscape, climate change in Kaata is fractal; change in each sphere occurs inter-relationally with changes in other spheres. Whether we talk of the crops or of human health, just as in Bastien's day there is similarity across scale; this node is connected to every other of the narrative.

IDELFONSO'S CLIMATE CHANGE WORKSHOPS: TRACING THE DISCOURSE

The term *cambio climático* ("climate change") was by all accounts introduced to the village by Idelfonso, a *campesino* from the neighboring village of Amarete, who led workshops on the theme in 2005 for the Ministry of the Environment. He is also the regional delegate for the left autonomist national indigenous social movement, the Conamaq. It is since he conducted these workshops, Idelfonso tells me, that the villagers talk about the "contaminating" effects of waste items such as plastics and batteries. The workshops took place five years before I arrived in Kaata, yet I was able to talk to Idelfonso about them, as well as to trace the discourse among the villagers.

The workshops seem to have focused largely on the communities' own waste, a theme they have taken up seriously. Idelfonso notes that "inorganic waste, plastic, batteries, all this is part of climate change. We aren't so much to blame in the communities' *cambio climático*—sometimes they consider us big contaminators." Laying the responsibility for climate change onto the *campesinos* presumably allows the Ministry to take robust charge of its "other," who though often idealized is considered somewhat inferior in education and understanding, especially in modern matters like rubbish, also distastefully inimical to contemporary idealizations of indigenous peoples as inhabiting a realm of "nature."

For Idelfonso, like the neighbor above, climate change is connected to networks extending beyond the village. Shifting upstream a little, this regional deputy who negotiates between the communities, social movements, and the national level has his own ideas about the source of the contamination. He tells me: "the wind comes not from oneself alone! The smoke comes not from oneself alone! The wind is moving... Climate change affects from far away." "From other countries?" I ask. "From multinationals, capitalists, extractivists, whatever you want to call them," he replies firmly. In his view climate change is caused by capitalism. "Seriously, from our own traditional methods we don't contaminate at all, no? This must be made clear, too." It is not Idelfonso's own belief that the "contamination" springs sheerly or even mainly from the villages themselves, yet as he is aware, this is the sense in which it is taken up by those residing in Kaata.

CONTAMINANTS AND THE WEATHER

The concept of contamination is taken very seriously by the people of Kaata, accustomed as they are to their actions being networked into an animate landscape, constantly constituting and being constituted by the soils, elements, and the spirits that inhabit it. They talk to the rains to draw them out of the sky or send them away, leave offerings of flowers filled with honey and alcohol in the crevices of the mountain to feed and encourage it to return crops. Within this complex network of relationships, my view, or Idelfonso's, that their actions were of minimal consequence in the landscape in contrast to those of a faraway set of consumers, would make little sense.

The following is an excerpt from a conversation I had in 2011 with a man with a ruddy complexion called Don Félix, at the time one of the authority holders in the village, and his friend Valentín. It took place one night over a bag of coca in the company of some of the other authority holders of that year. The conversation occurred in Spanish, his second language:

"There is contamination everywhere. Before there were no batteries, no plastics, there were none of these poisons either. So it was purely"—here he pauses, searching for an adequate word in Spanish—"purely natural," says Don Valentín.

"Natural! It was before. These things they're buying from the shops, they used to buy these in *uncuñas* before. They would weigh (the products) and put them straight into an *uncuña*."

"What's an *uncuña*?" I ask.

"It's woven, the size of a napkin—but woven though," he emphasizes. "Now, these are nylon! And these disposable drinks containers that come. These are contaminating us dreadfully, dreadfully. So I say that, as they say, this contamination affects the ozone layer of the sun, this is why the sun doesn't protect us like before. It falls, hell! Like it would burn us. Some burn [the consumer waste], this is dreadful. It would be good to bury them 40 meters underground. Buried rubbish is affecting us a lot too! There inside they are wearing out, exhausting more land. Even if you bury this plastic it does not rot. The batteries they throw away, dreadful!"

"Here people don't know this. Rather, they think this stuff is fertilizer! One who studies, [knows] it does not work. This is why they don't take care of it. Some of us now collect and bury those batteries. This Don Valentín," he jokes, poking fun at the man beside him, "he thinks it's fertilizer, he goes and collects it from the streets."

Félix's world converses with and engages the modern. His views on consumer waste resound with those of ecologists, thanks to the workshops Idelfonso gave. Searching for a way to express a state without this waste he is directed to the word "natural," taking up the modernist narrative of disjuncture between the fruits of human action or "culture" and "nature."

We note that the modern binary becomes necessary as a separating line across the landscape to talk about the contamination of consumer products. What exactly is different about this waste then compared with other substances? What exactly constitutes it as contamination? We will explore this over the course of this chapter; for now let us note that in the recent past Félix describes there were shops, but people would wrap their purchases in woven cloths rather than in contaminating plastics. For Félix, shopping as such does not constitute contamination, especially when it is for toasted grains from the community wrapped in a woven *uncuña*. I was told several times in other contexts that "cloth is culture." As I have explored elsewhere, woven wrappings are the product of a landscape, animate with local relationships and skills, and alive in a way that contaminating city waste or indeed city clothes are not (Bold 2017).

Félix's teasing of his companion Valentín indicates how consumer waste is seen in a context in which everything whether used up or left over is still involved in a circulatory household economy where nothing is wasted and no external inputs are required. Waste products are usually fertilizer; what is left over by the animals feeds the plants, and vice versa. Hence Valentín's conception, indeed a widespread conviction in Kaata, that batteries were fertilizer. The villagers organized a campaign to collect as many of them as they could following Idelfonso's workshops; prior to this many had ploughed them into their fields. Stobart (2006) recounts that on a journey he threw bean husks by the wayside, but noticed that his companions had kept theirs to take home and feed to the dogs. If the husks were left by the wayside, he was told, they would "cry." This waste is animate, such as children, animals, or crops, and if not tended to correctly by adults, it cries. While bean husks and woven cloths have their place in a system of relationships composing the landscape, plastics and batteries are "matter out of place" (Douglas 1966). Rather than fertilizing the earth as part of an interacting network of relationships, underground the contaminating substances "wear out" the soil and so strain the network of reciprocal relations. The full horror of these placeless artifacts, which have no use or destination after selling the commodities they encase, becomes evident when they are seen as emanating from and feeding into a system rather than as individual entities. Here the relational landscape can illuminate aspects of modern practice concealed from those who benefit from rubbish collectors and fetishize commodities as individual items enshrined in disposable wrappings.

While burying the waste contaminates the ground, burning it according to Félix damages the ozone layer. Félix observes this in relational terms, as the sun "burning them" more than before, which resounds with Andean narratives of worlds ending, as I shall explore below. Interestingly the matter of "protection" translates well into Western naturalist environmental discourse—the ozone layer is commonly said to "protect" us from the sun's rays, the environment we live in calibrated for our survival by the elements. We have indeed as Latour (1993) asserts never been modern; in the West these traces of animism enliven everyday ideas of the environment and soils, spoken about and indeed envisioned as agentive and beneficent entities, creating openings for cosmopolitical encounters.

AYNI AND THE PACHAKUTI

In a story widespread across the Andes and analyzed in detail by Abercrombie (1998) the *chullpas* ("ancestral beings") from another *pacha* ("era or cosmos") were burnt alive in their houses when Jesus Christ, the sun, rose for the first time in the east. With this came the end of the era of the *chullpas* and the beginning of Christian (so the people of this age are identified) time separated into days, nights, and seasons. This transformation is called in the Andes a *pachakuti* ("cosmic revolution"). *Pacha* is often translated as an age or epoch, yet according to Catherine Allen (2016: 000) this is "inadequate in conveying the human and non-human consciousness that inhabits this moment." *Kutiy*, according to Mannheim (1986), conveys the cyclical return of two parts exchanging via *ayni*, a system of like for like exchange among equals that characterizes, for example, working for one's fellow community members in the fields. *Pachakuti* was no longer used in Altiplanic communities by the 1970s, replaced with the Christian term Judgment Day, but with the difference that this did not convey a *final* ending (Bouysse-Cassagne and Harris 1987). A change in the elements, we learn, can signal a sudden epochal shift; the sun can suddenly change character making the world unfit to live in for the *chullpas*, who were accustomed to damp and dark conditions. This change, like those we are now experiencing, ushered in a new cosmos or *pacha*.

Chullpas, mummified remains of ancestors entombed in a crouching position, were engaged in community rituals until recently. David Llanos Layme (2005) describes a ceremony in Kaata in which they were mainly used to stop an excess of rain. When I asked in 2011 whether the villagers still carried out rituals to control the weather I was told that they had been sold, apparently to archeologists from Cusco, by the authorities appointed for the previous year and are now absent from their situation under the main square, the heart of the mountain body, from where they "dominated everything here before."

This episode exemplifies the changing nature of the Kaatan landscape, where money as a means of entering new networks of exchange becomes more important than the very heart of the mountain. It is as though one landscape, that of economic exploitation of the non-human elements as "natural resources," is superimposed like a tracing over another in which these non-humans are powerful actors with the capacity to create change. These worlding practices collide, slide across, and contest one another for meaning-making capacity in contemporary Kaata.

The sale of the ancestral *chullpas*, beings of another *pacha* ("world"), marks the decline of the reciprocal landscape of *mañay* and *ayni*—the heart having being sold from it. The sale of these remnants of a *pachakuti* is a striking indication of the current *pachakuti* under way in which the landscape is transformed from one of animate beings interrelating through *ayni*. By removing the heart of the mountain and selling and confining the mummies to a museum their animating power is lost; *ayni* is weakening in the landscape and the beings it animated are transformed into inanimate materials for human exploitation.

Some of the elaborate complex of "earth practices" (Harvey 2007) still take place, although with a much reduced number of participants since Bastien's day: the new *chakra* ("sowing site") each year, for example, is awoken with ceremonies of music and offerings that it reciprocates with the harvest. The villagers still carry out other weather rituals, and they work—nevertheless, sometimes, *sometimes*, Félix tells me, lowering his voice in a frank yet mighty admission, *en vano vamos* ("we go in vain") to beseech the elements for conditions conducive to crops. This weakening of the exchange relationships composing and managing the landscape is essentially what climate change comprises for the Kaatans.

AGRICULTURAL CHEMICALS AND CONTAMINATION

Continuing the conversation with Don Félix:

> "With respect to the diseases, they did not exist before. I don't know what has happened. The institutions, I think they come with just more poison. The cows, the camelids, needed no medicine before. Those ambitious ones up there [in the high pasture lands] want more wool! So what did they put—medicines, vaccination plans, other things. And this is where the illnesses come from, now they're getting worse and worse. It's the same in the production [of crops]. The ambitious ones want to grow potatoes this big," Félix says, gesturing a huge size. "It can be grown to the size it should grow to, no more!"
>
> "Don't exaggerate," agrees Don Valentín.
>
> "To this end," he continues, "They bring improved seeds as well. In these come the illnesses that afflict the plants also. I must disagree when they say they have to spray crops. This is less convenient. I am more than an agronomic engineer, and let this lot tell you if I'm lying! When we spray crops—there are creatures in the earth, which help the earth—they move it. When we spray them, we kill all this. The earth gets even harder. Some insects are good!" [murmurs of agreement from people around].
>
> "Let's say, there's a material under the ground, 'we'll come out in the dry season' [it says]. It hides itself under the earth. You can see it clearly. There are little holes, I think with these you see it clearly. These are real pores! This is how it breathes. So the earth doesn't have any breath at all, we are killing it worse, when we spray."

Félix embraces an approach that would resonate with a biodynamic or ecological Western perspective. This stems from his experience of agricultural "aids," the anti-contamination discourse introduced by Idelfonso, the Kaatan understanding of crops as sentient actors, and the mountain as an entity with skin. Kaatans have chosen to keep their rotational cycle instead of engaging with chemicals in a Faustian pact to shorten the time taken to rest the soil. The reluctance to spray crops and distrust of chemicals is widespread through the village, as indeed among many people who engage with the soil even in supposedly de-animated landscapes. Notwithstanding this spoken rejection of pesticides I have however observed several families using them. Ambition is associated with the cycle of contamination. Chemicals bring contamination, but what brings chemicals to be used is the ambitious desire for self-enrichment. Contamination occurs in a moral context connected to human greed.

Félix's above description of the breathing soil is coherent with the cosmos of the animate mountain body, yet might equally have been uttered by gardeners in England. Degnen (2009) shows how in northern England gardeners regularly speak about plants as having intentionality and sentience. Intersections between humans and plants "include bodily characteristics like bleeding, sleeping, and breathing" and "extend beyond physiology to include subjective states such as cleverness and insanity" (Degnen 2009: 163). There are reciprocal identifications between plants and human bodies, which, after Scott (1989), she describes as "interpretants" of one another (Degnen 2009: 164). Degnen draws upon Scheper-Hughes and Lock (1987), who themselves draw upon Bastien's (1985) work to explore the inter-relationality of mountain and body as an example of "the constant exchange of metaphors from body to nature and back to body again." Bastien's observations about Kaata are thus employed to draw out neglected aspects of Western practice, leading the way for vital cosmopolitical conversations.

I am interested in Félix's defiant phrase—"these are real pores!" he claims, as if such knowledge had been challenged as "not real"; a reflection of the declining discourse of the animate mountain, challenged through school education. Although Degnen's British gardeners may employ the same language of "breathing" soil or "protecting" vapors that keep the earth at the right temperature, most of them if challenged would say that this attribution of agency is "not real"; the "real" consists only in the connections between things that we can measure and science establish (Degnen 2009: 152). There is the modernist dichotomy of the "real" and "not real" waiting to classify plant lore, which Félix here defies. According to Degnen, prioritizing the admission that these intersections of humans and plants are "not real" is to "gloss too easily over sets of meaning in everyday gardening practice that merit much closer attention" (Degnen 2009: 152), unfairly prioritizing the naturalist perspective over everyday thought and practice; the day-to-day experiences of these gardeners leave space for cosmopolitical dialogue on climate change.

Soil science agrees that the holes created in the soil by insects allows air in and out, just as Félix describes. Here Félix challenges the hegemony of the scientific "real" with a Kaatan view in which the mountain body has pores in its soil skin, and we see that the two collide and collude. We could say that the worlds of Kaata and science here converge. The Kaatan discourse does not on the whole oppose but embraces the scientific one, applying an extra layer of animation to the elements it describes, sensu Evans-Pritchard (1937). Conceived as a living body the soil's need to breathe is easily understandable. The telluric spirits often fetishized by the city or "West" as the domain of the indigenous "other," as well as the capacity to care for and respect these as living beings, are already within "modern" landscapes, as indeed the modern use of pesticides is within those of this "other."

Pesticides and the Fractal Landscape

I am sitting with Idelfonso in a café on the Ceja, the "brow" of the city of La Paz, the edge of the valley in which the capital lies. Here the city opens up to the vast space of the Altiplano, and the glaciated mountaintops emerging out of it are suddenly visible. Sitting at a Formica table, over the noise of 1980s' rock and folkloric music, I ask Idelfonso what the villagers think about pesticides:

> "They precisely think they are negative, because in spraying you are also contaminating, as it were, with these chemicals. If you spray a lot you are contaminating the plant itself. Because we know it goes in the pores of the plants... it can get to its fruit, no? This is why we have said... we will produce organically... If the earth is sick, it's not good for you either, is it?" [Here he uses the word *maretando* ("nauseous" or "with its head spinning in circles like a drunken person").] "So ecologically, naturally, nothing is contaminated".

"So it's a cycle?" I ask.

"It's a cycle," he echoes. "Nothing is separated, nothing! When you are spraying, the wind is carrying the smell... to the earth itself."

"Before, there was a program from the European Community, there were engineers and everything, they brought chemicals, taught people how to put them on, how to spray the crops. People proved it was no good afterwards. Now they say no more. This was 10, no, 15 years ago now." [The villagers found the chemicals to have damaging effects] "When you put chemicals [on the earth], for two or three years it produces well, then afterwards doesn't produce any more—this they have proven, too!"

Idelfonso claims that the villagers have successfully, through employing empirical techniques, established the limitations of agricultural chemicals, despite the EC initiative. It seems that maintaining crop rotation continues to be more effective in making the soil fertile. While there is widespread agreement among moderns that such chemically induced intensive farming practices can exhaust the soil, and the United Nations makes proclamations to this effect,[2] modern agriculture nonetheless widely employs them, creating an interesting contrast between our empirical villagers, interpreting their results in terms of an animistic affiliation with the mountain, and moderns aware via science that they are wearing out soils, yet somehow unable to disengage from this cycle of contamination.

Idelfonso evokes a landscape in which humans, plants, and pesticides are interrelated in a "cycle," where the health of one element will mirror that of the others. While there is much change in this landscape, this cyclical reflection across scale seems familiar from the healing practices employed by the shamans Bastien (1978) described, where human bodies are healed through curing the mountain body. As Mannheim indicates, reciprocity is also cyclical: the verb *kutiy* conveying cyclicity or return. This landscape resembles the rhizome proposed by Deleuze and Guattari (1988), a network of interlacing threads in which every point is connected to every other, modeled on rhizomic roots like the potato, the staple Kaatan crop.[3] In Kaata climate change occurs within a network of actors such that if the contamination touches one "node," a human or crop, weakening it, it will likewise weaken the others.

There is sympathy between people, animals, and plants suffering the same changes at the same time. Chemicals have the same effect on plants and animals as on the people who eat them. Another man told me that we are these days like battery chickens growing rapidly but not strongly. Chemical feeds given to battery chickens to make them grow faster are thought to affect humans in the same way, accelerating the course of our lives rather than allowing us to come to maturity slowly.

CITY FOODS AND *AYNI*

Sitting over her loom in her front garden, Doña Carmen tells me:

> "Before we had potatoes this size," indicating a size greater than her hand. "Not now."
>   "Why not?" I ask.
>   "The earth is tired," she replies. "You have to plough it a lot. You plough in fertilizer, and it goes really far inside, and only stones come out. Only with a lot of fertilizer do things grow now."

The addition of lots of fertilizer was sometimes expressed by villagers as "obliging the earth to produce," really pushing their part of the exchange relationship with the soil. Other women reiterated Carmen's observation that the earth is tired, employing the word *pachamama* (literally "world mother") instead of earth. We note that the Kaatan narrative places agency with the earth—she is getting tired and sending stony infertile soil. This tiredness, as we have seen, can come from the contamination of rubbish, which "wears out the earth" beneath which it is buried. Chemical fertilizers are not considered a viable way of feeding the stony earth, indeed they tire "her" out. It is no wonder that harvests are diminishing.

It seems to my Western eyes that the Kaatan attribution of agency to the land conceals the human side of this relationship; from an anthropocentric viewpoint, we might say that lacking the energetic labor of young people, who have largely migrated away, soil quality declines and the work is harder. The young people desire fruit not of the soil. Devalued, unworked, unfed, and littered with the detritus of these desires the soil becomes weaker. Whereas a modern perspective projects a relation of causality from humans to non-agentive nature, here actors emerge in conjunction with one another as goods circulate in this landscape. Ingold (2006: 19) describes animism as a continual formation of beings—"an understanding of life as a creative process in which forms undergo continual generation, each in relation to the others."

The *productos* ("food crops") of the mountain emanate from and are rich in work and exchange relations with the earth and the elements. It is these *ayni* relationships that constitute entities or actors and make food-stuffs nutritious. City foods are considered by contrast to be less nutritious than those grown on the mountain. I was told that rice was not nutritious, unless it had been threshed by hand:

> They say that, before, there was not so much pasta or rice. Today people go to work to earn rice. There are some indigenous peoples who are from the same place where rice is produced. They thresh it, they don't use a machine to peel it. This is food, this threshed rice, but when it's done by machine, it's no use at all.

Rice threshed by machine is "no use at all," perhaps much as plastic wrappings could not nourish the soil. The food that people used to eat, which was the produce of the mountain, was more nutritious and made them stronger:

> Our produce, which they say we used to eat, had nothing of rice or pasta [murmurs of agreement from people around], nor did they add salt... People ate native things before, this is why the people of before were strong!

These assertions of Don Félix are coherent with those of the neighbor leaning over his wall that city-bought pasta and rice haven't got the same—vitamins. People are weaker and wearing out quicker as they consume city food, comprised of fewer complex relationships, which contribute an animating force as nutrition. Valued in terms of the labor that produced them the fruits of factories have little value. Although considering it a phenomenon unrelated to the labor conditions that produced the food, nutritionalists would again agree that these foods are indeed less nutritious, another collision or "resonance" (Prigogine and Stengers 1984) of the two sets of invested meanings or universes. We can agree on some things in the pluriverse, apparently, although our methods of constructing meaning differ.

Weaker People

It is generally affirmed that contemporary people are weaker and less hardworking than the "people of before":

Before, people used to get up early in the morning—our ancestors, the people of before. We are become lazy now. We have a *mink'a* ['a communal work day, in which participants are paid by the owner of the fields'], let's say, today we're going to plant wheat. The *mink'as* ['the daily workers'] are arriving at ten, half-ten, they're chewing their coca, at eleven they start working. They work an hour, they eat lunch, there's no progress with the work. Worse. They say that, before, people were mature, they were more capable than we are today. We are failing in everything now.

These days people do not exert themselves to work for one another as they did before. Gose (1994: 113) explains how paid *mink'a* work expresses hierarchical relations, whereas *ayni* is key to creating networks of reciprocity between equals, creating a "powerful sense of moral connection." Gose (1994: 113) describes large *ayni* work parties, and notes that one man defined *ayni* as "having the tenderness to work for others." In contemporary Kaata I have only seen *ayni* carried out between small groups, usually extended families. The community is increasingly rent by inequality, and the former egalitarian exchange relationships are declining. This failure to work hard is what the man cited above summarizes as a lack of maturity, a concept that surfaces repeatedly.

Henry Stobart (2006) describes the Andean life cycle as separated into two halves: that of youth, characterized by "crying," need, and dependency, and that of maturity, characterized by catering to the needs and cries of others. Indeed it seems maturity is constituted through feeding rather than crying to be fed (Stobart 2006). It is through fulfilling these relationships of nurture that actors become mature. People now, as described above, are not seen as mature since they do not tend a landscape of dependents as they once did; to merely consume is to remain a child. Like contamination or health, maturity is also a sympathetic attribute of humans and their animals and products.

It is largely the young people through their consumption of city-made foods who cause waste. Their appetite for sweets and drinking yogurt, which is sold in sachets and is one of the major processed foods present in Kaata, was commented on with concern. As another man remarked: "Today our children don't want indigenous (*originario*) produce. They are used to what comes from outside. Pasta... various things from there, isn't it, they value them. That's it! They like them." This desire, the changing value attached to the consumption of goods by young people, is key to this cycle of contamination and climate change, as they are led away from the mountain and its values. The resonance this has for the community, already having to contend with contamination, is mass migration, which is inseparable from the earth being "tired" and stony, the fiestas being small, and from there being no one to play music.

Music and Batteries

Continuing my conversation with Doña Carmen:

> There are fiestas which, more or less, we performed with rituals... This we are forgetting now, this 'culture,' we say. Totally, totally now. These days the kids don't know how to play anything! They only listen to the radio— *aiii*, my ears! They'll drive themselves crazy. They don't know how to play anything. This is the culture that we are forgetting, day by day. It's at the end now. I think it's ending this class of before, of those who played. For our *pachamama*, for our... production, there were fiestas.

Music and fiestas are key to defining the vanishing "people of before." Ideas about music, as Harry Stobart indicates:

> Are not neatly separated from other spheres, but are deeply integrated into general ideas about 'production.' Through rituals, you appeal to places and objects and set them in 'communicative mode' through offerings and referring to them in affective ways, house as nest, fields as virgin, potato as *chaskanawi*. Playing music and dancing also open these lines of communication, where well- being and reproductive potential are understood as linked with the quality of relations. (Stobart 2006: 5)

We might say that through music one can call the landscape into inter-relationality and a particular way of being. Among the common purchases of the young migrant workers are radios, which until the village was connected to the national grid a few years ago, required batteries to power them. This substitutes for the traditional music of the community, the drums and long flutes, their tone and tunes varying with each fiesta. Idelfonso's workshops seem to have struck a resonant chord with the village in describing these batteries as contaminants. They map a landscape of networks toward a new culture coming from the cities, rather than networks in which traditional music awakens the fields or calls the rains. Climate change is a discourse of the weakening of the land and the people on it.

Winds of Change

To conclude the story I told at the start of the chapter, I asked my neighbor during our conversation across his wall if schemes providing paid employment for young people within traditional food production would allow the community to stop climate change. He threw up his arms and laughed, "No, you can't stop it! The contamination is borne in the winds, from other places, from big factories, you can't stop it"—he looked at me and tried to explain—"you can't stop a car polluting, for example. And here we are contaminating too…"

In Kaata the wind is a living entity: Ina Rösing (2003, 2010) notes that the wind is addressed as *Ankari*, is key in most rituals as it carries "effects" to the spirits or persons to be entreated and cured, and is celebrated as a deity. This animate wind, which can carry "effects" from afar, is similar to what we might think of in the West as "atmosphere" in the artistic sense as well as in the scientific sense in terms of the vapors surrounding the earth, which we humans inhale and exhale. The air in the Andes is suffused with moods as well as animate and living presences. As Ingold (2006) indicates, when we cease to regard the world as surfaces and see it instead as a constantly shifting medium through which we move, as from an animist perspective, then we will realize the air is the primary medium in which we exist.

Climate change has something of the texture of a mood that animates everything: a mood of contamination that sweeps through the landscape, sickening and weakening the lines of relationships connecting it, borne in the winds or atmosphere. The atmosphere! Once again here we see a convergence with science, for it is indeed in a scientific sense that this very atmosphere is the carrier of contamination. The animistic perspective does not hinder but complements science; the two universes or ontologies collide with one another and we see we are talking about the same element: the very air with which human beings are constantly interacting and the same contamination whether in terms of named gases and particles or as a mood-carrying substance. As with the breathing earth the animistic view is so literally true in our scientific universe that it challenges the hegemony of scientific ways of understanding, expanding material into its affective aspects, and un-splitting the scientific with its confining causality into a total phenomenon encompassing moral dimensions reflecting directly on human actions. And it is precisely these aspects that scientific climate analyses lack.

Pachakuti

Don Félix states categorically that climate change only comes from local sources of contamination; for him it is a complex of changes embedded in the landscape of the mountain. The neighbor leaning over his fence, who speaks good Spanish and has worked in tourism, connects climate change to factories, which are indeed the provenance of non-nutritious foods and contaminating wrappers. It is the same set of economic distribution networks that are implicated across scale, whether we talk of the immediate spatial landscape of the village or carry it further to the factories that are their source.

Idelfonso's is a self-consciously anti-capitalist narrative. "With these plastics, there we are contaminating," he concedes. "As indigenous peoples, we have always respected the earth—*¡el suelo nosotros respetamos!*" he claims, in a tone of perfect humility and reason. "We are part of the mother earth!" he asserts, "Some people say it [the earth] has to depend on us, but this is the wrong way round! We are *part* of the mother earth." Latour (2017: 14) identifies such affirmations of belonging to the earth as "seemingly simple expression[s] that [are] actually obscure," at least to a modern audience. Idelfonso engages with the modernist discourse that would allocate humans control over nature, thus separating them from it, and challenges it with an affirmation of continuity with nature. He highlights the difference between humans taking care of the earth, as *causa eficens* in an inanimate nature, or being part of it, as *causa naturans*, or one actor among an animated multitude (Heidegger 1971). In Idelfonso's narrative it is capitalism that is the contaminating force; his views align with Moore's attribution of agency in the Capitalocene (Moore 2015).

The hegemony of capitalist modernity among governments forecloses the identification of consumerism or "greed" and the utilitarian evaluations of value in the environment as root causes of climate change, continuing to prevent sensible actions to address it. The dimensionality of a world configured by consumption hides such realities from our eyes, in much the same way that the packaging in which consumer products are encased are rendered invisible through effective waste disposal systems. In this sense Idelfonso's or the Kaatan view are more complete, holistically connecting human behavior and attitudes and their environmental consequences, than the modernist capitalist perspective that pulls up short of connecting these related phenomena in the popular or mainstream imaginary.

Marcelino Yanahuaya told Bastien that the life-and-death cycles of society, the village, above and below, were like those of the fields, the fallow plots drawn into life again by the Qallay rituals to awaken them. "Time is not a set point, but a cycle between two strokes always circulating within the mountain's... bodies. Like the swing of a pendulum, each stroke can only go so far, and then it starts back again. Furthermore, each stroke propels the other..." (Bastien 1978: 53). Gose (1994: 137) agrees with this metaphor, noting that "the principle of *ayni* does inform the parallel contributions of the living and dead to agriculture, and the cyclical movements of sun and water between their respective worlds, which Bastien (1978: 53) aptly compares to the stroke of a pendulum." Related in these cycles of counterbalancing alterity, then, we can read the capitalist transformation of the village and the young people's appetite for Western customs and clothes, as well as the contamination and coming cataclysm that inseparably accompany them, which may bring it to ruin.

Some people in Kaata say that when the snows of the glaciers, now rapidly diminishing as temperatures increase, eventually disappear the mountains they cover will explode into volcanoes and bury this world beneath them. Such cataclysmic disasters, it is said, will happen sooner in capitalist places and eventually spread to Kaata. Villagers cited contemporary events, such as the Japanese tsunami and Chilean earthquakes, as evidence that this process was already well under way, clearly understanding that the environments of more "developed" countries are also relational and subject to violent checks by non-humans. We may view this as the swinging of the pendulum: as our exchange relationships with the earth and mountains—animate non-humans in the landscape—are stretched taut by humans taking too much the balance of exchange will reverberate and spring back as the mountain engulfs and absorbs us back into itself. Oblitas Poblete, holder of a small hacienda in Charazani and author of a well-researched account of Callawaya culture, explains that opposing dualistic forces constitute the universe, which in their most elemental form are forces of maintenance and destruction; when the destructive force predominates the world is destroyed by hurricanes and earthquakes (Oblitas Poblete 1963).

The explosion of the mountains resembles a classic *pachakuti*. Classen (1990) indicates, drawing on the seventeenth century writings of chronicler Garcilaso de la Vega, that the *pachakuti* is a period of sacred and highly dangerous fluidity between the land and body during which humans emerge from the natural world and can also return to it (1990).

Kuti we remember describes a "turning over" and, according to Bruce Mannheim, implies cyclical return as well as the alternation of two elements exchanging via *ayni* (Mannheim 1986). *Pachakuti* has a sense of cyclicity, as Mannheim indicates, involving the cycling of reciprocal energy between humans and earth. Indeed such a world ending has happened many times before—at the time of the conquest chroniclers recorded nine previous *pachakuti* in Incaic narratives of the past.

We might compare this to the Incan narrative of the start of their *pacha*. Viracocha, the serpent god, found humans hiding in caves on its surface and by teaching them to worship the mountains allowed them to practice agriculture, thus inscribing a new order on the surface of the earth (Classen 1990). This is a pattern of surface and interior occupying a similar spatial logic to the volcanic *pachakuti*: as humans come out from caves onto the surface of the world a new existence emerges once again from the midst of the previous one; worlds are born from one another. The end of the world here is not the end of humans *nor* of the world (cf. Danowski and de Castro 2017), but their *reversal* as agency shifts, and what was once below revolves now to be above.

Conclusion

The *pachakuti* with its connotations of cyclical return is illuminating in a context where *anthropos* has come to take too much and strain its relationships with the "environment." We can usefully conceptualize climate change as a swinging back of the pendulum or, as Latour (2017: 80) put it, the "revenge of Gaia" on the modern view that the world is simply "natural resources" for human consumption, a.k.a. the modern de-animating ontology, and its imaginary Anthropocene of control over "nature." This is a major way in which these perceptions of climate change from Kaata are useful for a modern audience, or in which we might fruitfully enter into a cosmopolitics (Stengers 2005) of climate change.

We moderns can also learn from the idea of maturity by caring for the world as a system and by being aware of the cyclical connectivity across species crucial in creating the landscapes we can see and conditions for human life. Greed and ambition are linked to contamination in Kaata; the health of the ecosystem can be maintained by not taking too much. The relational view of the world allows us to conceptualize the close inter-relationality and resemblance across scale of human bodies and the landscape where environmental and bodily pollution are inseparable.

Although the Kaatans perceive themselves to be deeply implicated in climate change, as noted in the conversation with my neighbor, they consider that they cannot stop it. Their embrace of the role played by each fragment of litter might be held up as an environmentally responsible indigenous ideal, yet—and this is the sting in the tail of the animist perspective for a Western ecologist—the landscape of dispersed agency that they describe is not one in which human actions can necessarily save the whole. We are dwarfed by the agency of animate mountains and subsumed by the flow of currents of reciprocity. The animate landscape does not supply a solution to the issues created by the Anthropocene, the secondary effects of environmental fallout, for in respecting and animating the multitude of actors of "the environment," humanity ceases to have agency over them. There is no magic solution from the "other," the indigenous or "people outside capitalism" that it might have been hoped would provide a timely balm for resolution of environmental destruction accompanying extractive capitalism. I hope to have shown here, however, that there are some things we can learn from considering climate change cosmopolitically, if we are in time to practice them.

NOTES

1. Andean word connoting a village community and its lands.
2. https://www.scientificamerican.com/article/only-60-years-of-farming-left-if-soil-degradation-continues/.
3. Technically a potato plant is a tuber, which is structurally a rhizome, with swollen fleshy growths.

REFERENCES

Allen, Catherine J. 2016. "The Living Ones: Miniatures and Animation in the Andes." *Journal of Anthropological Research* 72 (4): 416–441.

Bastien, Joseph W. 1978. *Mountain of the Condor: Metaphor and Ritual in an Andean Ayllu, American Ethnological Society Monographs*. St. Paul, MN: West Publishing.

Bastien, Joseph W. 1985. "Qollahuaya-Andean Body Concepts: A Topographical-Hydraulic Model of Physiology." *American Anthropologist* 87 (3): 595–611.

Bessire, Lucas, and David Bond. 2014. "Ontological Anthropology and the Deferral of Critique." *American Ethnologist* 41 (3): 440–456. https://doi.org/10.1111/amet.12083.

Bold, Rosalyn. 2017. "Vivir Bien: A Study in Alterity." *Journal of Latin American and Caribbean Ethnic Studies* 12 (2): 113–132.

Bouysse-Cassagne, Thérèse, and Olivia Harris. 1987. "Pacha: En torno al pensamiento Aymara." In *Tres reflecciones sobre el Pensamiento, Andino*, edited by Olivia Harris, Thérèse Bouysse-Cassagne, Tristan Platt, and Veronica Cereceda, 11–57. La Paz: Hisbol.

Classen, Constance. 1990. *Inca Cosmology and the Human Body*. Montreal: Faculty of Religious Studies, McGill University.

Danowski, Deborah, and Eduardo Batalha Viveiros de Castro. 2017. *The Ends of the World*. Cambridge: Polity Press.

Degnen, Cathrine. 2009. "On Vegetable Love: Gardening, Plants, and People in the North of England." *Journal of the Royal Anthropological Institute* 15 (1): 151–167. https://doi.org/10.1111/j.1467-9655.2008.01535.x.

Deleuze, Gilles, and Félix Guattari. 1988. *A Thousand Plateaus: Capitalism and Schizophrenia*. London: Athlone Press.

Douglas, Mary. 1966. *Purity and Danger: An Analysis of Concepts of Pollution and Taboo*. London: Routledge.

Evans-Pritchard, E. E. 1937. *Witchcraft, Oracles and Magic Among the Azande*. London: Oxford University Press.

Gose, Peter. 1994. *Deathly Waters and Hungry Mountains*. Toronto: University of Toronto Press.

Harvey, Penelope. 2007. "Civilising Modern Practices: Response to Isabelle Stengers." Paper given at the meeting of the American Anthropological Association, Washington, DC.

Heidegger, Martin. 1971. *Poetry, Language, Thought*. New York: Harper & Row.

Holbraad, Martin, and Morten Axel Pedersen. 2017. *The Ontological Turn: An Anthropological Exposition, New Departures in Anthropology*. Cambridge: Cambridge University Press.

Ingold, Timothy. 2000. *The Perception of the Environment: Essays in Livelihood, Dwelling, and Skill*. London: Routledge.

Ingold, Timothy. 2006. "Rethinking the Animate, Reanimating Thought." *Ethnos* 71 (1): 9–20. https://doi.org/10.1080/00141840600603111.

Latour, Bruno. 1993. *We Have Never Been Modern*. Cambridge, MA: Harvard University Press.

Latour, Bruno. 2017. *Facing Gaia: Eight Lectures on the New Climatic Regime*. Cambridge, UK and Medford, MA: Polity Press. https://ebookcentral.proquest.com/lib/duke/detail.action?docID=4926426.

Llanos Layme, David. 2005. "Ritos para detener a la lluvia en una communidad de Charazani." In *Gracias a Dios y a los achachilas. Ensayos de sociología de la religión en los Andes*, edited by A. Spedding, 159–184. La Paz: Plural-ISEAT.

Mannheim, Bruce. 1986. "The Language of Reciprocity in Southern Peruvian Quechua." *Anthropological Linguistics* 28 (3): 267–273.

Moore, Jason. 2015. *Capitalism in the Web of Life*. New York: Verso.

Oblitas Poblete, Enrique. 1963. *Cultura Callawaya*. La Paz: Achivo y Biblioteca Nacionales de Bolivia.

Prigogine, Ilya, and Isabelle Stengers. 1984. *Order Out of Chaos: Man's New Dialogue with Nature*. New York: Bantam Books.

Rösing, Ina. 2003. *Religión, ritual y vida cotidiana en los Andes: los diez géneros de Amarete: segundo Ciclo Ankari: rituales colectivos en la región Kallawaya, Bolivia*. Madrid: Iberoamericana.

Rösing, Ina. 2010. *White, Grey and Black Kallawaya Healing Rituals*. Madrid: Iberoamericana.

Scheper-Hughes, Nancy, and Margaret Lock. 1987. "The Mindful Body: A Prolegomenon to Future Work in Medical Anthropology." *Medical Anthropology Quarterly* 1 (1): 6–41.

Scott, Christopher. 1989. "Knowledge Construction Among the Cree Hunters: Metaphors and Literal Understanding." *Journal de la Société des Américanistes* 75 (1): 193–208.

Stengers, Isabelle. 2005. "The Cosmopolitical Proposal." In *Making Things Public: Atmospheres of Democracy*, edited by Bruno Latour and Peter Weibel, 994–1004. Cambridge: MIT Press.

Stobart, Harry. 2006. *Music and the Poetics of Production in the Bolivian Andes*. Aldershot, Hants, UK: Ashgate.

Shifting Strategies: The Myth of Wanamei and the Amazon Indigenous REDD+ Programme in Madre de Dios, Peru

Chantelle Murtagh

INTRODUCTION

Indigenous peoples in the Amazonian regions of Peru have been vocal in their criticism of some recent global climate change initiatives including the UN REDD+ (Reducing Emissions from Deforestation and Degradation) Programme. REDD+ is aimed at reducing greenhouse gas emissions from deforestation and degradation. Leaders from

This chapter is based on arguments that I develop in the chapter "Producing knowledge and constituting power" of my PhD thesis titled "Producing leaders. An ethnography of an indigenous organisation in the Peruvian Amazon" (see Murtagh 2016). Sections from the thesis chapter are reproduced here.

Research for the PhD was undertaken in the Madre de Dios region of southeastern Peru during 2012–2013 and was possible thanks to a NWDTC studentship funded by the Economic and Social Research Council in the UK [Grant No. 1091812].

C. Murtagh (✉)
University of Manchester, Manchester, UK

© The Author(s) 2019
R. Bold (ed.), *Indigenous Perceptions of the End of the World*,
Palgrave Studies in Anthropology of Sustainability,
https://doi.org/10.1007/978-3-030-13860-8_6

the Native Federation of the Madre de Dios River and Tributaries (FENAMAD) highlight the significance of their "cosmovision" or worldview in developing the Amazon Indigenous REDD+ Programme (REDD+ Indígena Amazónica—RIA). This is an alternative REDD+ programme developed and promoted by indigenous organizations in Latin America that aims to address some of the needs of indigenous peoples while contributing to the reduction in deforestation. An Amazon Indigenous REDD+ pilot project was under proposal in the Amarakaeri Communal Reserve (RCA) in the Madre de Dios region at the time of my doctoral fieldwork. In this chapter I will explore how this cosmovision influences the Harakmbut people's understanding of climate change and its solutions, undertaking the analysis through a discussion of the myth of Wanamei.

The myth of Wanamei is a prominent Harakmbut myth, as recounted by Harakmbut leaders from the Madre de Dios region of Peru. This area is located in the southeastern Peruvian Amazon, bordering Brazil and Bolivia. The Wanamei myth tells of a time of climate change resulting in widespread destruction through a great fire and a flood. It also talks of salvation. Human action is highlighted as capable of influencing the outcome of events, and we are taught that attention needs to be given to the role of the spirit world in human affairs. Indeed, there are entities in the spirit world who are able to aid humans, and maintaining correct relationships with them is essential for human existence. Through an analysis of the myth we are able to see the world as multi-layered, including both visible and ordinarily non-visible components, each interacting with each other to influence states of being and outcomes, including the weather. Seeing the world in this manner draws our attention to a holistic worldview Harakmbut leaders state is lacking in present climate change discussions. Additionally, I draw on the myth to focus attention on the way in which indigenous understandings of history and change have influenced their interactions with and modifications to the UN REDD+ Programme.

REDD+ was developed by parties to the UN Framework Convention for Climate Change as a way of mitigating climate change by reducing "emissions from deforestation and environment degradation," initially through placing a financial value on carbon stores held in intact forests. It has since expanded and developed to include more of an emphasis on co-benefits including poverty alleviation, improving livelihoods, and forest governance[1] (see Howell 2017). The general idea is that private companies and developed nations assist developing countries and local stakeholders to continue to conserve their forests through payments for environmental services, carbon funds, and trading carbon stocks through financial markets.

I will draw attention to indigenous organizations' responses to the UN REDD+ Programme, exploring how the Amazon Indigenous REDD+ emerged as a response to conflicts of representation indigenous leaders experienced. This counter-programme was felt to address the criticisms laid out by indigenous leaders and offer a viable solution for indigenous communities interested in participating under the various REDD-type initiatives, which also include payment for ecosystem services. I focus on the discussions in 2011 and 2012 surrounding REDD+ coming out of the RCA, exploring how indigenous understandings of their world and the past play a significant role in shaping present actions. In the second section of the chapter I focus on the myth of Wanamei, using the notion of "historicities" to highlight the cultural significance of remembering the past (Whitehead 2003a; for other examples that highlight the significance of remembering the past see High 2015[2] and Walker, chapter "Fragile Time: The Redemptive Force of the Urarina Apocalypse" in this volume). Whitehead (2003a: xi) has suggested that "historicities" are the

> culturally constituted texts, visual and aural representations, verbal narratives and oral and somatic performances that are the discrete tales that make specific histories [...] By historicities is meant the cultural proclivities that lead to certain kinds of historical consciousness within which such histories are meaningful.

I argue that the myth of Wanamei provides Harakmbut leaders with a framework to understand the past and present (see Gray 1996, 1997a, b). Hugh-Jones (1989: 54) suggests that "despite being the essence of timeless tradition, myth is still subject to a constant process of change, which allows it to keep pace with reality." The Wanamei myth describes a change in climate where human action is instrumental. The myth allows for lessons to be learned to mitigate problems. Harakmbut leaders today refer to this myth in their descriptions of the significance of the RCA. It acts as a reminder for them of the interconnectedness of all beings (humans, animals, plants, and spirits) and is used in discussions on REDD+ to highlight their holistic worldview.

Harakmbut understandings of territory are influenced by what might be termed their cosmovision, including material and symbolic ties to the space, which Escobar (2016) terms "thinking-feeling with the earth." This holistic cosmovision includes an understanding of lived environment as composed of dynamic and interactive social relationships between humans, plants, animals, and spirit beings (Gray 1996, 1997a; Moore 2003; Surrallés and García Hierro (eds.) 2005b; Descola 2005). For the Harakmbut, like other Amazonian peoples, humans are among many agentive actors interacting in the environment. This is also a view which is shared by the Asheninka people and Kaata residents as described by Comandulli and Bold in chapters "A Territory to Sustain the World(s): From Local Awareness and Practice to the Global Crisis" and "Contamination, Climate Change, and Cosmopolitical Resonance in Kaata, Bolivia", respectively, of this volume. For the Harakmbut people an imbalance in the correct relationships between these different entities and beings can have negative effects on humans and cause changes to the environment including the climate. Surrallés and García Hierro suggest that Amazonian indigenous peoples understand their territories as "relational space" and the "consolidation of a very specific and singular fabric of social ties between the different beings that make up that environment" (Surrallés and García Hierro 2005a: 11). Descola (2005) describes this in terms of animistic ecologies based on a cosmology in which animals and plants form part of what he refers to as a "community of persons." They are considered to "share most of the faculties, behaviours, and moral codes ordinarily granted to humans" (Descola 2005: 24). This allows us to reflect on the type of social relations that can be formed between these different beings. Escobar (2016) describes such worldviews as "relational ontologies," where "nothing pre-exists the relations that constitute them." These are worlds in the making, constantly brought into being through actions (Escobar 2016; Ingold 2006; Ingold 2001). This relational space is therefore dynamic, interconnected, and changeable. This holistic and interrelated way of viewing the world comes into conflict with worldviews promoted through programs such as REDD+, which aims at dissecting the components of "nature" and the lived environment, placing financial value on some aspects of the natural world over others.

I argue that this cosmovision has played a part in shaping the Harakmbut leaders' responses to the REDD+ Programme and led them to seek to influence its development and implementation. I end the chapter with a discussion of the shift in power dynamics resulting from greater involvement of indigenous people in decision making through their interactions with the Amazon Indigenous REDD+ Programme. Greater indigenous involvement in international political spheres such as those set out by the United Nations present the possibility for bringing new worlds into being.

Ethnographic Context

I conducted PhD fieldwork between 2012 and 2013 in the Amazonian region of Madre de Dios in southeastern Peru. I spent much of my time working with indigenous leaders at the FENAMAD, which has its office in the city of Puerto Maldonado. The Federation is a regional indigenous organization representing 34 native communities in the Madre de Dios River Basin from seven different ethnic groups, including the Amahuaca, Ese Eja, Harakmbut, Yine, Matsiguenka, Kichua Runa, and Shipibo Conibo. At the time of my fieldwork there were a number of Harakmbut leaders who had been elected to the Federation and it is to this ethnic group I refer in this chapter.

The Harakmbut people are an Amerindian people whose ancestral lands are in the lowland Amazonian regions of Madre de Dios and Cusco. There are a number of different Harakmbut-speaking groups, of which the Arakmbut, formerly known as the Amarakaeri (Gray 1996: xvii), are now the most populous. Contact with disease as a result of the rubber boom in the early twentieth century decimated indigenous populations in the region (Gray 1996). Harakmbut groups were later contacted by Dominican Missionaries in the 1940s and 1950s. Some families spent a brief time in the Shintuya Mission located on the upper Madre de Dios River before going on to form their own communities (Gray 1996, 1997a). The Madre de Dios region has changed dramatically in recent years due to surfacing of the inter-oceanic highway linking the coasts of Peru and Brazil (see Harvey and Knox 2012). The consequences of this has been an influx of migrants from the Peruvian highlands and marked increases in illegal gold mining and illegal logging activities. The region is otherwise known as the "Capital of Biodiversity of Peru."

REDD+ in Peru and the Emergence
of the Amazon Indigenous REDD+ Programme

The REDD+ or "Reducing Emissions from Deforestation and Forest Degradation in Developing Countries" initiative aims to provide financial incentives for the protection of forests to mitigate the effects of climate change. Carbon stores are calculated and given a financial value that can then be traded. The idea is that local stakeholders are encouraged to see standing trees as more valuable than timber and are provided with financial rewards for their role as protectors of the forest.

The reality of initiating the programme in Peru, however, was rather more complicated. There were high start-up costs for developing the projects and especially for carrying out the logistical and background studies needed to quantify the carbon stores (see Brightman and Lewis 2017: 8 for a similar discussion). These steps created the need for intermediaries. These intermediaries were often NGOs, leading to suspicion over who would benefit from the financial rewards offered by the programme. It was unclear how these rewards would be distributed to local stakeholders (also discussed in Conservation International 2012). Initially there were minimal state regulations and interventions, which meant that a number of private companies, as well as national and international NGOs, approached communities directly. In some cases 100-year contracts were offered, which were thought to complicate the issue of resource ownership and use over time, especially due to the fact that many indigenous communities were still in the process of petitioning for official land rights.

The initial response from indigenous organizations was to oppose the programme completely and prohibit NGOs and private companies from entering communities with project proposals. In 2011 a meeting of regional indigenous organizations was called by The Interethnic Association for the Development of the Peruvian Rainforest (AIDESEP). AIDESEP is the national indigenous organization of Peru to which most of the regional organizations are affiliated. The results of the meeting were formulated in the Iquitos Declaration (2011).[3] In this document the indigenous leaders present agreed that carbon markets did not offer a viable response to the climate problem and indeed served to exacerbate local issues in relation to land tenure and resource rights (see Larson et al. 2013; Sunderlin et al. 2018). The following excerpt from the Iquitos Declaration highlights some of their concerns:

> We reject that instead of recognising and paying their historical ecological debt, global powers keep on polluting, deforesting and pillaging, and intend to cover this up through the "carbon market", through contracts with companies that falsely "compensate" such damage with payments to indigenous communities and local people for conserving their forests which clean up this contamination. Contracts which serve to accentuate the loss of control over our ancestral lands, forms of life and rights, which will be traded on the stock exchange with huge profits. It is unacceptable to risk the suicide of our forms of life by insisting on more and more business (Iquitos Declaration 2011; *my translation*).[4]

Around the same time indigenous communities saw the emergence of what have been called "carbon cowboys." These NGOs or private companies aimed to tap into the financial rewards offered by developing projects with indigenous communities who, at that point, had little knowledge about what the programme was. The best documented example, mentioned in the Iquitos Declaration, comes from the Matses people who were allegedly persuaded by a private Australian company to sign a contract in English giving over their rights to their forests in exchange for a single payment of US$10,000[5] (see also Sarmiento Barletti and Larson 2017).

The REDD+ Programme therefore came to be seen as a potential danger to indigenous territorial rights and was prioritized in the indigenous agendas of both the regional and national organizations. In the 2012 Congress of FENAMAD a presentation about REDD+ was given by COICA (Coordinator of Indigenous Organizations of the Amazon River Basin, an international indigenous-run organization with a base in Ecuador). This gave indigenous people the opportunity to discuss their experiences of the REDD+ Programme. A general consensus established that more information was needed about the programme before communities could make informed decisions about how to engage successfully with it. Klaus Quique, who was the vice-president of FENAMAD said at the time "ever since 2008, REDD has become fashionable. It's all about carbon. The NGOs have besieged the indigenous communities. But first we have to get to know the topic of REDD." Antonio, the president of one of the Arakmbut communities, added "there is no clarity in the communities about REDD. However, there are contracts where there is no clear information [...] They coordinate with the communities in order to be allowed in and then tell the communities that they are doing environmental services. FENAMAD needs to open up workshops about this and the NGOs should approach FENAMAD first" (for an analysis of REDD+ in the context of indigenous rights and abuses see Sarmiento Barletti and Larson 2017).

The lack of clarity with regard to REDD+ led indigenous organizations to call for communities to refrain from entering into discussion and negotiation with NGOs or private companies. Some communities in Madre de Dios, however, went on to accuse the Federation of trying to monopolize the financial rewards of this programme. These communities wanted access to the financial rewards the NGOs or companies were offering and considered that their involvement with the REDD+ Programme could provide them with the means to invest in their own futures. The Federation took the position that REDD+ was in danger of violating important indigenous rights such as the right to land security and control of resources. They also asserted that it did not present a viable solution to the climate problem. After another meeting about REDD+ with indigenous peoples in FENAMAD's office a statement was published on FENAMAD's Facebook page (February 25, 2013) claiming that "indigenous peoples are worried by the commercialization of the environment and the control of lands and resources. We denounce that this mechanism violates our rights and could have impacts on the governing structures of our indigenous peoples and communities who depend on the forests" (*author's translation*[6]).

The pressure to find a solution to the problems with which REDD+ presented the indigenous communities was being felt by the federations and their leaders. In particular it became necessary to find a way to avoid alienating communities who wanted to actively engage with the programme. There was a gradual shift away from blanket refusal of all interactions with REDD+ toward a process of discussion and negotiation. The organizations started to further emphasize that "we cannot talk about REDD+ without taking into account the autonomy, self-determination and judicial guarantees of indigenous peoples' territory" (also included in the Iquitos Declaration 2011[7]). AIDESEP and FENAMAD stated that "without land titles there is no REDD+."[8] In addition, the significant role of NGOs and other intermediaries was highlighted to demand that negotiations about the terms of contracts and payments be entered into directly with local actors[9] (see also Conservation International-Peru 2012: 73).

It was out of these continued discussions between national and regional federations that the Amazon Indigenous REDD+ (RIA) Programme was born. The programme was developed to convert what was considered to be the threat of the REDD+ Programme into a significant and viable opportunity for indigenous people to further their own development.

The RCA was chosen as the site for a pilot Amazon Indigenous REDD+ project, developed by COICA and AIDESEP in collaboration with FENAMAD. The aim was to approach the forest conservation programme in a new way through emphasizing the holistic value of the forests and shifting focus away from their role as carbon stores. The indigenous leaders wanted to put a value on the whole territory including the plants, animals, water, and people as well as trees. Written into the programme are considerations for territorial security and indigenous rights. There is an attempt to ensure possibilities for the integral management of the land by looking at ways to integrate the environment, culture, and society, a view which the leaders considered to be more in line with indigenous cosmovision (see AIDESEP article: More than carbon, its life and culture, http://www.aidesep.org.pe/node/12736). Cosmovision here implies viewing the world in terms of the intricate interconnections between all living things as well as animal and spirit beings. This makes it difficult to divide the environment into separate parts, some of which are given more value than others. This was one of the motivations for pushing for more holistic management options to mitigate or adapt to climate change.

AMARAKAERI COMMUNAL RESERVE

The RCA covers an area of 402,335 hectares, officially recognized as a protected area by the state in 2002 (Decreto Supremo No. 031-2002). The indigenous peoples of Madre de Dios were active in pushing for this recognition, a struggle that lasted many years. The area covered by the reserve forms part of what the Harakmbut peoples consider to be their ancestral lands. As a specific category of protected area the communal reserve is co-managed by the state and 10 native communities[10] made up of Matsiguenka, Yine, and Harakmbut villages, all of which share their boundaries with the reserve. The Executor of Administration Contract or ECA, an elected executive committee with representatives from the communities, works closely with the state's National Parks Service (Servicio Nacional de Áreas Naturales Protegidas por el Estado, SERNANP) via the Amarakaeri Reserve Management Office. Together they oversee the protection and management of the area. Recently the Arakmbut people have been working on creating detailed maps of the area in an attempt to document places of cultural, archeological, mythological, and historical significance.

This work was given greater impetus due to the presence of an oil-drilling concession, block 76, designated by the state in 2006. The petrol concession, which is currently owned by Hunt Oil, overlaps the RCA almost entirely and prospecting work has already been carried out.[11] As well as oil exploration there are major problems in the reserve and surrounding areas due to illegal logging and gold mining, increasing the territorial insecurity felt by native communities. Mining is a major sector contributing significantly to the Peruvian economy, promoted by national government through development and infrastructure projects. The regional government has also played a significant role in promoting the mining sector in recent years.

Despite these conflicting agendas the centrality of this area for the Harakmbut people is reflected in the framing of one leader's description of its significance. Fermin Chimatani, an ex-president of ECA, commented in an interview published online:

> The RCA is the territory of the Harakmbut peoples: it is the territory where Wanamei was born, the mythological and sacred tree which saved the Harakmbut people from floods and fire millions of years ago, which gave us new hope for life. This territory is threatened by various activities resulting from man's greed, putting our existence at risk by endangering our water, our biological and cultural resources. Our myth, according to the ancestors, testifies that Wanamei appeared to save my people from this threat.
>
> It is our home, where the spirits of our ancestors are; our history, our life, our reason for being is there. [...] For SERNANP it is a Natural Protected Area, created to achieve a specific conservation objective, which in some ways is a view shared by indigenous people, and that is why we signed an indefinite management contract with them. For the oil company, however, it is an area whose only use is for the extraction of oil and gas and to gain money and then leave, and for this reason they are not interested if the communities continue to live" (*author's translation*; https://arsenico. lamula.pe/2014/10/08/entrevista-a-ferminchimatani/siriusblack1603/ date first accessed April 22, 2016).

Gray suggests that: "When myth is performed, a historical act brings the timeless past into contemporary relevance" (Gray 1996: 199; see also Gow 2001; Rappaport 1998; High 2015; Whitehead 2003b). We can therefore read the myth of Wanamei as a reflection of contemporary human experience, and as such its meaning lies in the relevance of its message for the Arakmbut people today. Gray suggests that the myth produces a world for the Arakmbut people that only takes on meaning when listeners recognize its message (Gray 1996: 220; see also Hugh-Jones 1989). The recounting of the story of Wanamei provides us with an insight into how aspects of this myth come to frame the Harakmbut engagement with the state and the REDD+ Programme.[12]

THE MYTH OF WANAMEI

The myth of Wanamei[13] tells of a time when there was a great fire, when everything was destroyed: the animals and plants were burned and even the fish were found floating dead, having been boiled in the water. The sky was filled with smoke and people began to believe that the whole jungle would be engulfed. The dreamers or visionary leaders, called *wayorokaeri*, who can see what is happening in other places through their dreams, alerted the people (see artistic depiction of this in Fig. 1). While they were meeting to discuss things one of the *wayorokaeri* noticed a parrot flying overhead holding a seed. The *wayorokaeri* said, "I dreamt of this parrot, who has come to help us to escape the fire—we need to give him a woman." They presented the parrot with various women of different ages, none of whom seemed to be suitable, for the parrot did not give over his seed. Eventually he favored a young virgin girl. The seed was placed between her legs and a tree grew quickly. The dreamers asked the tree to help them as the fire was now surrounding their village. The tree lowered its branches so that the people could climb up it to safety. All the animals also climbed up the tree. The people in the tree did not eat any animals and the animals did not bite the people. The people looked after the tree as it was their protector and provided refuge, food, and fruit for them to eat.

After the fire stopped, intense rains came, which caused widespread flooding. There was a period of darkness too. The rains stopped after some months and the sun came out, but the earth had been converted into a great swamp that swallowed up anything that fell into it. The men fired arrows into the land, testing the firmness of the ground, and these

Fig. 1 A section from the myth of Wanamei painted on the wall of the Santa Cruz church close to the main market in the city of Puerto Maldonado, Madre de Dios, Peru. The picture was painted by Harakmbut children as part of the Equilibrium Festival in November 2015. (Translation of words from Spanish: Over time a disorder among humans caused the sun god to announce through the dream of the *curaca* ["a leader or wise man"] that there would be a great drought and a purifying fire with the possibility of being saved) (*Photo credit* Author)

disappeared in the mud. For days, weeks, and months the people lived in the tree that had saved their lives, until one day someone fired an arrow into the earth and its end remained sticking out. After some more time they fired another arrow into the ground and it bounced back. The people descended to the ground but disappeared into the mud as the ground was only firm close to the tree. This left only two people, a brother and sister, who were forced to have relations with each other to populate the world, and it is said that this is the reason why humans die (for a complete version of this myth see Gray 1996, 1997a; sections of the myth are also discussed in Moore 2003).

Animals are considered by the Arakmbut to have given men knowledge via the visions of the *wayorokaeri*—*sabios* ("those that know")— who receive their information in dreams, by restricting themselves to special diets, inhaling *rapé* tobacco or through the use of ayahuasca (*Banisteriopsis caapi*). Animals have powers due to the *wanokireng* ("spirits") they have in their bodies (Moore 2003: 78), thought to be spirits of human ancestors who became incorporated into animals when they died. As a result, these spirits are able to help their kin. The *curanderos* ("healers or shamans") are called *wandakaeri* and are also able to communicate with the spirit world through visions and songs.

In an extended version of the myth the surviving couple request the help of the woodpecker to acquire fire to cook food as they are hungry. The woodpecker steals fire from a dangerous spirit called Toto in the invisible world. Toto is the owner of fire and guards its secrets. The woodpecker creates a distraction by telling Toto about the terrible fire–flood and takes the opportunity to steal the fire stick. It is through stealing fire from the invisible world that the woodpecker makes it available to the visible world and to human beings. Here we are drawn to the fact that there are both spirits who are willing to help humans and others who can harm them and often there are tensions between them (Gray 1996: 37–38). Both the parrot and the woodpecker in the Wanamei myth are able to mediate between the sky and the earth, essentially the visible and invisible worlds. The myth draws our attention to the importance of the invisible world as often problems that arise in the visible world can be resolved through consultation with the spirit world. It can be seen that these channels of communication are essential for humanity's existence (Gray 1996: 38).

Discussion of the Myth as a Particular Way of Viewing the World

In the beginning according to the myth there was chaos in the world. People ate everything without care or distinction, and had indiscriminate sexual relations, even with siblings (Moore 2003: 76). For the Arakmbut maintaining order in the cosmos requires doing the same in the natural lived world. There are rules that emphasize that care should be taken with regard to which animals can be hunted and eaten and by whom. Sharing, abundance, and generosity are looked upon favorably and care should be taken not to overexploit natural resources (Gray 1997a: 119).

Incest is prohibited and the Harakmbut groups are exogamous. Divergence from these social expectations could risk offending agents in the spirit world and may disrupt the cosmic order (Moore 2003: 79).

The Wanamei tree provided shelter for the Harakmbut people in the mythic past and allowed for a liminal period of learning and reflection while protecting them from the chaos surrounding them. The myth also serves to highlight the importance of communication with the spirit world for human existence, as exemplified by fire being given to humans by the woodpecker, or the parrot who had the Wanamei tree's seed (Gray 1996: 28–29). The myth highlights that attention should be paid to the interconnections between humans and their lived environment, including the relationships that are nurtured with spiritual beings and animal and plant protectors. Only in this way can order be maintained in the cosmos. For the Arakmbut most problems also have a supernatural component and often answers can be found by recourse to spirit entities through the work of the *wayorokaeri* ("dreamers") or *wandakaeri* (*curanderos* or "shamans") who can communicate with the invisible world (Moore 2003: 73). The Harakmbut consider mythic explanations for many events and natural phenomena, and the supernatural is therefore considered to be "an extension of the natural order as it serves to broaden and order reality" (Moore 2003: 62).

The Harakmbut people and the other indigenous peoples in the region have been active in working with their federation, FENAMAD, to take actions to protect the RCA as a dynamic, vibrant, and living resource for future generations. They assert that protection furnishes a means for the continued survival of their people.

Focusing on the tree in the myth of Wanamei allows us to reflect on the way in which it has become a potent symbol for the construction of the collective identity of the Harakmbut people. It provides a central origin narrative as the different Harakmbut groups and clans were formed once they had come down from the tree (Gray 1996; see also Whitehead 2003b: 61, 76 for a discussion on symbolism). Once the humans had descended the tree then disappeared into the ground. Some Harakmbut people describe that the tree is located at the headwaters of the Madre de Dios River, an area which is now located within the RCA. This adds weight to their attempts to protect and conserve this area, especially since it is said that the tree will once again rise up to save the Harakmbut people at the time of the end of the world (Gray 1996: 28).

Both the tree and the reserve therefore act as powerful symbols used to establish a sense of place.

More broadly the myth is about a transition from a catastrophe to better times and from chaos to order. People were suspended in a liminal state in the tree, and after danger subsided they were able to return to their ordinary lives, so the myth tells of salvation and hope for a better future. "In the myth, creation is the synthesis or resolution of extremes by means of a go-between or mediator" (Gray 1996: 30). Parallels can be drawn here, as indigenous leaders take on the role of mediator between indigenous peoples and others, speaking for and on behalf of the communities and working to create order and establish conditions favorable for living well. By engaging with the state in the co-management of the communal reserve, a form of order can be said to have been established. This means that the Harakmbut people are able to maintain some level of use and control of the area and can play an important role in decision making with regard to resource management, which will benefit all the communities. The leaders continue to push for the right to self-determination: to be able to decide how they want to live, this being one of the main pillars of the indigenous movement (Gray 1997b; Jackson and Warren 2005). To be able to live well and maintain relations with the entities that ensure this possibility the Harakmbut people need their territory. Territoriality is, therefore, related to the material and symbolic ties to the space (Escobar 2017). The myth helps us to see the interconnections between the beings that inhabit this space.

The Harakmbut people see time as cyclical, oscillating between dry and wet seasons, new and full moon, day and night, and these cycles determine daily activities (Gray 1996: 61–65). According to Gray this view extends also to their interpretations of history and myth, as both are seen to involve transitions between day and night. He argues that in their history there have been moments of chaos and of tranquillity where they were able to live well. The myth of Wanamei forecasts that a time will come when the tree will reappear to save the Harakmbut people again, and events will repeat themselves, demonstrating the cyclical nature of the myth (Gray 1996: 65). Interestingly, Permanto (chapter "The End of Days: Climate Change, Mythistory, and Cosmological Notions of Regeneration" in this volume) also talks of repeating mythistories. The myth is used by the Harakmbut people to make sense of both their present and past experiences and can be revisited to frame interactions and decisions in the present (Whitehead 2003b). For Terence Turner mythic and historical consciousness are often complementary ways of framing the same events. He emphasizes "the dynamic role of historical consciousness as the repository of alternative courses of action in the present, which may […] become decisive ingredients for present action" (Turner 1988: 212).

It is through the Amazon Indigenous REDD+ Programme that we can see how indigenous peoples are actively reshaping their lived world in terms of their understanding of relational landscapes. The role of indigenous people in its development and plans for implementation has been significant. This was what Klaus Quique was referring to when he said "They always talk to us about pilots [pilot projects] but we never have the plane" (FENAMAD meeting, February 21, 2013). He felt that the Amazon Indigenous REDD+ counter-programme would offer an opportunity to rectify this situation. Over the last few years indigenous leaders have promoted the Amazon Indigenous REDD+ Programme at climate change conferences including the UN Climate Change Conference held in Paris in 2015 (Figs. 2 and 3).

Fig. 2   UN Climate Change Conference in Paris. Indigenous leaders (from left) Fermin Chimatani and Jaime Corisepa (representatives from the Executor of Administration Contract—RCA) and representatives from COICA (*Photo credit* Ricardo Burgos, November 30, 2015)

Fig. 3 Poster for side events organized by indigenous organizations at the UN Climate Change Conference in Paris (*Photo credit* Poster published and widely distributed to promote side event at COP)

SHIFTING STRATEGIES BASED ON DIFFERENT WORLDVIEWS

One morning I went to the FENAMAD offices early to talk to Luis Tayori, a young politically active Arakmbut leader, about his experience of climate change. Over a number of years he has worked diligently collecting life histories and cultural and historical information about the Harakmbut people and has coordinated a number of field trips to the RCA to document and produce maps about their ancestral lands. That day he was sitting at his desk translating interviews from Harakmbut into Spanish. When I asked about the REDD+ Programme and the Amazon Indigenous REDD+ Programme he stressed the importance of indigenous peoples being involved in these discussions. He said, "why can't indigenous people be involved if this is the language used at a national and international level? We modify the use of the words with our indigenous vision. So, we want to put a value not only on carbon but on every organism, even ants." Then he smiled and added, "It's a different *pollera* but the same *chola.*"

Pollera refers to a specific type of traditional skirt worn by highland indigenous women in Peru, Bolivia, and Ecuador, mostly Quechua speakers; *chola* is used to refer to these same "citified indians." In Peru the phrase is a common colloquial expression used to mean that the same basic reality is being presented under a new guise and often implied is a motive of deception or cynical repackaging. I was left wondering why he would choose to phrase his engagement with the REDD+ Programme in these terms.

From the perspective of indigenous people the REDD+ Programme could be seen as a new type of economic activity, following closely on from other extractive industries that had served to appropriate indigenous resources and marginalize indigenous peoples' involvement in their resource management. In this way REDD+, by placing a financial value on carbon as opposed to wood, oil, or gas, was felt to represent the next step in the commercialization of the forests. Leaders felt that participation in the programme could result in territorial insecurity for indigenous peoples. Many communities had seen an increase in infighting and divisions not dissimilar to what they had experienced through their involvement with other types of extractive activities. They were assigned an essentially passive role in the programme as "protectors of the forest" but did not have any clear input into the design, planning, or management (Howell 2017). They felt that their role was further limited by

NGO involvement. In addition, it appeared that they would potentially receive only limited benefits from conservation. In many respects it was felt that this programme allowed business to continue as usual without addressing the main causes of climate change, offering only a new mechanism for financial gain from the environment. It could be argued then that the REDD+ Programme was a different *pollera* that masked the same *chola* reality of the business agenda and natural resource exploitation, offering indigenous peoples limited possibilities for change.[14]

It is my opinion, however, that Luis was using the phrase to highlight the agency of indigenous people in shaping their own futures. For him the Amazon Indigenous REDD+ Programme was the new skirt and the indigenous people and their agendas the *cholas*. The Amazon Indigenous REDD+ Programme sought to upset the balance of power by affirming the indigenous peoples' role in the design and management of any such programme. By using and adapting the new mechanisms introduced by the state and the international community, albeit with a critical perspective, indigenous peoples are able to further their same demands for territorial consolidation, autonomy, and self-determination. Changing strategies, or putting on a different skirt, allowed them to continue their political struggle for visibility and control of their lands and resources.

The *chola* has often implied a specific type of power relation of domination and subordination. Weismantel describes how the *cholas*, as market selling women, are racialized and sexualised. However, the *pollera* can also be seen as a symbol of female power and financial independence (Weismantel 2001: xxv; xxvii; see also Seligmann 1989). It can be argued that the view of the *chola* is changing in recent years with the increased participation of indigenous people in decision-making processes and in asserting their rights. This has resulted in renewed pride in indigenous identity.[15] I posit that the Amazon Indigenous REDD+ Programme offers the opportunity to subvert the categories of exclusion and re-balance the power dynamic by allowing indigenous peoples the possibility of becoming at least co-pilots of the plane.

Change is needed for new worlds to be born. Indigenous peoples, like the Harakmbut, may not be surprised by changes in climate as theirs is a world in flux, constantly in the making and dependent on the cultivation of correct relationships with others. This involves living well together through displays of generosity and sharing, and cultivating relations with significant plant, animal, and spirit beings. Climate change has happened before according to their myth and could easily happen again if there is disorder in human affairs and the relations that con-stitute a landscape are "out of sync" due to inattention to the correct ways of "being in the world" (Ingold 2001). The new emphasis that the international community, through the REDD+ Programme, has placed on carbon appears to allow for business to continue as usual, without addressing what the Harakmbut consider to be the underlying problem: a disconnect between humans and the other entities that make up the lived environment.

Indigenous people, by ensuring that they maintain control of their lands and their autonomy in decision making, feel that they will be bet-ter able to manage the effects of the changes to come. The Amazon Indigenous REDD+ Programme also focuses on the need for a life plan for adaptation to climate change. This plan involves the low-impact man-agement of the communal use of forest resources based on ancestral knowledge systems, stating that people and forests are one. It rejects the emphasis on carbon markets and focuses rather on ecosystem services as a whole.

The Harakmbut people believe that as Wanamei saved them once it can do so again if they are able to maintain their links to the spirit world and the correct relations that they have with these entities. Having access to their land will ensure that these links can be maintained into the future. So, climate change can be understood as the "end of the world" in a number of senses. In one sense the world (life and land) as we know it may change as a result of the variations in climate, but in another the focus on climate change has resulted in the end of *a* world through a readjustment of the power imbalances that have historically marginalized indigenous peoples. Indigenous peoples are now able to assert their posi-tion and voice their demands at an international level and have been able to modify climate change strategies. New worlds are brought into being and new possibilities are opening up.

NOTES

1. For a more complete discussion of different components of the REDD+ Programme it may be useful to refer to the research being undertaken by the Center for International Forestry Research (CIFOR) and in particular their Global Comparative Study on REDD+. Since 2008 they have been studying REDD+ in different contexts across the globe and have pro-duced a number of significant publications on the topic (https://www.cifor.org/gcs/).

2. High (2015) discusses the role of social memory in how the Waorani define their relationships, both with each other and in relation to others.
3. Iquitos Declaration (2011). See http://www.redd-monitor.org/wp-content/uploads/2011/05/1371.pdf (accessed June 20, 2016).
4. Iquitos Declaration (2011). See http://www.redd-monitor.org/wp-content/uploads/2011/05/1371.pdf (accessed June 20, 2016).
5. For more information about the specific Peruvian case see http://www.redd-monitor.org/2012/09/18/judge-in-peru-issues-warrant-for-carbon-cowboy-david-nilssons-arrest/. For more information on the emergence of carbon cowboys around the world see http://www.redd-monitor.org/?s=carbon+cowboy (accessed June 20, 2016).
6. A los pueblos indígenas les preocupa la comercialización de la naturaleza y el control de las tierras y los recursos naturales. Denuncian que este mecanismo viola sus derechos y que podría tener impactos sobre las estructuras de gobierno de los pueblos indígenas y comunidades dependientes de los bosques (https://www.facebook.com/FENAMAD/photos/a.207420545958322.58444.201180499915660/536972293003144/?type=3, accessed August 14, 2018).
7. Iquitos Declaration (http://www.redd-monitor.org/wp-content/uploads/2011/05/1371.pdf) and AIDESEP pronouncement of October 28, 2010 (https://www.servindi.org/actualidad/34447, accessed July 23, 2017).
8. http://aidesep.org.pe/index.php/noticias/pueblos-indigenas-de-madre-de-dios-reafirman-su-compromiso-sin-titulacion-no-habra (accessed June 10, 2018). See also the No REDD platform (https://redd-monitor.org/2011/09/22/no-redd-platform-issues-wakeup-call-to-funders/, accessed August 4, 2018).
9. Extract taken from the *Plataforma de Lucha*, published in one of the pamphlets printed during the regional strikes in Puerto Maldonado in 2009.
10. The "native community" is a specific legal category in Peruvian law that allows indigenous peoples the possibility of being granted land titles to lands they have occupied ancestrally or they currently use (Decreto Ley No. 22175, http://www2.congreso.gob.pe/sicr/cendocbib/con3_uibd.nsf/0D41EC1170BDE30A052578F70059D913/$FILE/(1)leydeco-munidadesnativasley22175.pdf).
11. As of June 2015 Hunt Oil has suspended explorations in the RCA. See Hill (2015). US's Hunt Oil suspends drilling in Peruvian Amazon (http://www.huffingtonpost.co.uk/david-hill/uss-hunt-oil-suspends-dri_b_7561416.html, accessed May 2, 2017).
12. See Gow (2001) for an interesting discussion on myth and history. I think that mythic understandings of the past become incorporated into how

some indigenous people and particularly indigenous leaders talk about themselves and in relation to others. This is especially relevant in contexts of uneven power relations such as those between the national state and indigenous peoples. The past then becomes a powerful tool used to legitimize indigenous claims to recognition of their rights in the same way that cosmovision and culture can become reified for the same purposes (Conklin and Graham 1995; Conklin 1997). It is through their interactions with the state and others that indigenous people can be seen to be "actively debating and revising the meaning of their own culture" (Turner 1991: 308; see also Oakdale 2004). Rappaport (1998: 15) suggests that history can "become a vehicle for empowerment" and I think that mythology can also serve the same purpose. See Hugh-Jones (1989) for a divergent view on the use of myth to legitimize political claims.

13. Wanamei is also written as Wanamey. Andrew Gray, for example, uses Wanamey in his book series. I decided to use Wanamei as opposed to Wanamey as it was most commonly used by the leaders I worked with in the Native Federation of Madre de Dios (FENAMAD). Interestingly, at the time of my fieldwork, collaborative work was being undertaken with the Peruvian Ministry for Education to develop a standardized version of the Harakmbut language to make materials in Harakmbut available to schools in the different Harakmbut communities. These workshops incited debate due to the different variations of the language used by the various Harakmbut-speaking groups in the region.

14. Some indigenous groups see both REDD+ and Indigenous REDD+ as being *polleras* that cover the basic *chola* reality of capitalist exploitation of the forest. See, for example, http://www.movimientos.org/es/madretierra/show_text.php3%3Fkey%3D21905 (accessed November 10, 2016).

15. For example, the BBC article: The rise of the "cholitas" (http://www.bbc.co.uk/news/magazine-26172313, accessed April 4, 2017).

REFERENCES

Brightman, Marc, and Jerome Lewis, eds. 2017. "Introduction: The Anthropology of Sustainability: Beyond Development and Progress." In *The Anthropology of Sustainability: Beyond Development and Progress*, edited by Marc Brightman and Jerome Lewis, 1–34. New York: Palgrave Macmillan.

Conklin, Beth. 1997. "Body Paint, Feathers and VCRs: Aesthetics and Authenticity in Amazonia Activism." *American Ethnologist* 24 (4): 711–737.

Conklin, Beth, and Laura Graham. 1995. "The Shifting Middle Ground: Amazonian Indians and Eco-Politics." *American Anthropologist* 97 (4): 695–710.

Conservation International. 2012. Una aproximación a la participación plena y efectiva en las iniciativas REDD+ en Peru. https://www.conservation.org/global/peru/publicaciones/Documents/documento_participacion.pdf. Date Accessed 6 April 2017.

Descola, Philippe. 2005. "Ecology as Cosmological Analysis." In *The Land Within: Indigenous Territory and Perception of the Environment*, edited by Alexandre Surrallés and Pedro García Hierro, 22–35. Copenhagen: IWGIA.

Escobar, Arturo. 2016. "Thinking-Feeling With the Earth: Territorial Struggles and the Ontological Dimension of the Epistemologies of the South." *Revista de Antropología Iberoamericana* 11 (1): 11–32. Madrid: Antropólogos Iberoamericanos en Red. http://www.aibr.org/antropologia/netesp/numeros/1101/110102e.pdf.

Escobar, Arturo. 2017. "Sustaining the Pluriverse: The Political Ontology of Territorial Struggles in Latin America." In *The Anthropology of Sustainability: Beyond Development and Progress*, edited by Marc Brightman and Jerome Lewis, Ch. 14, 237–256. New York: Palgrave Macmillan.

Gow, Peter. 2001. *An Amazonian Myth and Its History*. Oxford: Oxford University Press.

Gray, Andrew. 1996. *The Arakmbut: Mythology, Spirituality and History in an Amazonian Community*. Oxford: Berghahn Books.

Gray, Andrew. 1997a. *The Last Shaman—Change in an Amazonian Community: The Arakmbut of Amazonian Peru*. Oxford: Berghahn Books.

Gray, Andrew. 1997b. *Indigenous Rights and Development: Self-Determination in an Amazonian Community*. Oxford: Berghahn Books.

Harvey, Penny, and Hannah Knox. 2012. "The Enchantments of Infrastructure." *Mobilities* 7 (4): 521–536.

High, Casey. 2015. *Victims and Warriors: Violence, History and Memory in Amazonia*. Urbana: University of Illinois Press.

Howell, Signe. 2017. "Different Knowledge Regimes and Some Consequences for 'Sustainability'." In *The Anthropology of Sustainability: Beyond Development and Progress*, edited by Marc Brightman and Jerome Lewis, 127–144. New York: Palgrave Macmillan.

Hugh-Jones, Stephen. 1989. "Wãrĩbi and the White Men: History and Myth in Northwest Amazonia." In *History and Ethnicity*, edited by Elizabeth Tonkin, Maryon McDonald, and Malcolm Chapman, Ch. 4, 53–70. London: Routledge.

Ingold, Tim. 2001. *The Perception of the Environment: Essays on Livelihood, Dwelling and Skill*, Ch 1, pp 13–26. London: Routledge.

Ingold, Tim. 2006. "Rethinking the Animate, Re-Animating Thought." *Ethnos* 71 (1): 9–20. https://doi.org/10.1080/00141840600603111.

Jackson, Jean, and Kay Warren. 2005. "Indigenous Movements in Latin America, 1992–2004: Controversies, Ironies, New Directions." *Annual Review of Anthropology* 34: 549–573.

Larson, Anne, Maria Brockhaus, William Sunderlin, Amy Duchelle, Andrea Babon, Therese Dokken, Thu Thuy Pham, Galia Selaya, Ida Resosudarmo, Abdon Awono, and Thu-Ba Huynh. 2013. "Land Tenure and REDD+ : The Good, the Bad and the Ugly." *Global Environmental Change* 23: 678–689. http://www.umb.no/statisk/ior/land_tenure_and_redd.pdf.

Moore, Thomas. 2003. "La etnografía tradicional arakmbut y la minería aurífera." In *Los pueblos indígenas de Madre de Dios historia, etnografía y coyuntura*, edited by Beatriz Huertas Castillo and Alfredo García Altamirano, 18–35. Lima Perú: IWGIA.

Murtagh, Chantelle. 2016. "Producing Leaders: An Ethnography of an Indigenous Organisation in the Peruvian Amazon." PhD thesis, University of Manchester.

Oakdale, Suzanne. 2004. "The Culture-Conscious Brazilian Indian: Representing and Reworking Indianess in Kayabi Political Discourse." *American Ethnologist* 31 (1): 60–75.

Rappaport, Joanne. 1998. "Introduction: Interpreting the Past." In *The Politics of Memory: Native Historical Interpretations in the Colombia Andes*, 1–30. Durham: Duke University Press.

Sarmiento Barletti, Juan Pablo and Anne Larson. 2017. Rights Abuse Allegations in the Context of REDD+ Readiness and Implementation: A Preliminary Review and Proposal for Moving Forward. Center for International Forestry Research (CIFOR) Infobrief No. 190. https://doi.org/10.17528/cifor/006630, http://www.cifor.org/publications/pdf_files/infobrief/6630-infobrief.pdf.

Seligmann, Linda. 1989. "To Be in Between: The Cholas as Market Women." *Comparative Studies of Society and History* 31 (4): 694–721.

Sunderlin, William, Claudio de Sassi, Erin Sills, Amy Duchelle, Anne Larson, Ida Resosudarmo, Abdon Awono, Demetrius Kweka, and Thu-Ba Huynh. 2018. "Creating an Appropriate Tenure Foundation for REDD+: The Record to Date and Prospects for the Future." *World Development* 106: 376–392.

Surrallés, Alexandre, and Pedro García Hierro, eds. 2005a. "Introduction." In *The Land Within. Indigenous Territory and Perception of the Environment*, edited by Alexandre Surrallés and García Hierro Pedro, 8–21. Copenhagen: IWGIA.

Surrallés, Alexandre, and Pedro García Hierro, eds. 2005b. *The Land Within: Indigenous Territory and Perception of the Environment*. Copenhagen: IWGIA.

Turner, Terence. 1988. "History, Myth and Social Consciousness Among the Kayapo of Central Brazil." In *Rethinking History and Myth: Indigenous South American Perspectives on the Past*, edited by J. Hill, 195–213. Urbana: University of Illinois Press.

Turner, Terence. 1991. "Representing, Resisting, Rethinking: Historical Transformation of Kayapo Culture and Anthropological Consciousness." In *Colonial Situations: Essays on Contextualisation of Ethnographic Knowledge*, edited by George W. Stocking Jr., 285–313. Wisconsin: University of Wisconsin Press.

Weismantel, Mary. 2001. *Cholas and Pishtacos: Stories of Race and Sex in the Andes.* Chicago: University of Chicago Press.

Whitehead, Neil, ed. 2003a. *Histories and Historicities in Amazonia.* Lincoln: University of Nebraska Press.

Whitehead, Neil, ed. 2003b. "Three Patamuna Trees: Landscape and History in the Guyana Highlands." In *Histories and Historicities in Amazonia,* 59–80. Lincoln: University of Nebraska Press.

A Territory to Sustain the World(s): From Local Awareness and Practice to the Global Crisis

Carolina Comandulli

INTRODUCTION

> During a meeting I attended in Denmark with scientists I had a vision: I saw the world half frozen and half very hot. In the heat, I saw children that were adapting. The forest was being transformed. The adults wanted to run away, but there was a big hole where people would go and die. The earth was divided [...] I saw that if the Ashaninka territory is not looked after, half of the earth will fall. [...] We have an objective that the great spirit left for us: if you look after what is given to you, you will be looked after. We need to take care of everything that was given to. (Moisés, January 2016)

This chapter is about social renewal through the construction of a space of autonomy, inspired by shamanic vision. The Ashaninka from Amônia River have seen their world practically end. Despite the serious threats they have faced from patrons, loggers, and more recently climatic

C. Comandulli (✉)
Department of Anthropology, University College London, London, UK

© The Author(s) 2019
R. Bold (ed.), *Indigenous Perceptions of the End of the World*,
Palgrave Studies in Anthropology of Sustainability,
https://doi.org/10.1007/978-3-030-13860-8_7

Fig. 1 Moisés' drawing, showing their land and the destruction happening outside

changes, they have ingeniously constructed strategies to strengthen their own worlding practices, and are currently not only thriving but also willing to be an example to other indigenous and non-indigenous peoples in sustaining our shared world. The transformations and reconstruction the Ashaninka have been able to realize arise from a history of struggle against exploitative systems and in support of other-than-human beings that dwell in their lived space. The guidance of their shamans,[1] as well as obtaining a territory where their visions and thoughts could be materialized, have been key in their struggle.

In this chapter I outline important aspects of the "Ashaninka world" and contextualize the Ashaninka from Amônia River's struggle to prevent its destruction. I will describe the ongoing threats and challenges they face, and explore how they interpret this predatory human action. Finally, I examine contemporary shamanic thought and community action in response to the challenges of maintaining their world, showing how the Ashaninka are trying to reverse a state of destruction and engage other indigenous and non-indigenous peoples in this effort. This chapter derives from four years of research about the Ashaninka people, including 25 months of fieldwork with the Ashaninka from Amônia River between 2015 and 2017.

> We cannot forget where we live and who we are. If we forget our language and our traditions, we are forgetting to live the truth of being an Ashaninka. We are living in another world—a world of illusion. It is time we stop and reflect on how we want to live. We need leaders with pure hearts. [...] We need to go back to our world. What we are living in many Ashaninka communities is another world. (Moisés, August 2015)

The Ashaninka are an Arawak-speaking people who inhabit the Peruvian and Brazilian Amazon rainforest. They number more than 100,000[2] and are probably the biggest indigenous population of lowland Amazonia. In Brazil they live in Acre State and inhabit six indigenous lands (Ricardo and Ricardo 2017: 543–544). In Peru 357 titled native communities are spread over eight departments The total titled area the Ashaninka people inhabit covers nearly 3 million hectares of the Amazon rainforest (Piyako et al. 2015).

Since European colonization the Ashaninka people have endured more than four centuries of contact and violent challenges to their territories and lives by missionaries, timber patrons, colonization, and internal wars. In spite of the remarkable historical resistance they have shown they still face many challenges. Their lands in the Amazon are surrounded and targeted by extractive industries. In Peru the particularities of their land-titling process are provoking noticeable changes in their ways of living, as their titled lands are small and the communities are adopting a national schooling system which is deeply nationalist and does not value indigenous ways of being. Changes in climate are also affecting the environment and the predictability of local natural cycles, threatening their food sovereignty and consequently their health and survival.

In the face of such challenges and threats the Ashaninka from Amônia River present us with a compelling history of resistance and ingenuity on how to create a space of autonomy for their continuity. They have successfully fought for ownership of their territory; as Escobar states, territorial struggles are ontological struggles, as they "interrupt the globalizing project of fitting many worlds into one" (2017: 239). By defending their territory the Ashaninka from Apiwtxa have been able to sustain the existence of their own world. Not only that, they have managed to share their practices with other indigenous and non-indigenous communities, with the aim of looking after a shared space. As Escobar affirms, the knowledges produced in such struggles "might be particularly relevant for the search for post-capitalist, sustainable plural models of life" (Escobar 2017), highlighting the importance of such learning as a source of inspiration for other peoples in a time of crisis.

> To tell you a little bit of our history, the Ashaninka people have been massacred by loggers, we have been massacred by rubber dealers, we have been massacred by colonizers who came and killed our animals, we were taken as workforce to serve patrons who told us to cut down the forest and get the animals for them so that they could live well, we were massacred by the missions who told us that we knew nothing. But then we decided to give a different response: we began to study.
>
> My grandfather, Samuel Piyãko, was a great student, because he escaped the massacres where thousands of Ashaninka died, and he searched for an escape in distant places to be able to survive with nature: with pure oxygen, pure water, all natural. But then, when he crossed the mountains, he met again outsiders who were entering the rivers and invading. Then he said: "I do not have where to escape. I will have to adapt here. I will stay here and look with my spirit to see how I will be able to remain connected." So, he began analyzing everything around him: what was happening with the rubber tappers, with the peasants, with the loggers, and he developed a thought: "I will do things differently! I will look after my culture and look after the families that are under patrons' rule. And I will choose a wife for my son." As he did not know how to speak Portuguese, he chose a white woman, who is my mother, and told her: "You will be my son's wife. I had a spiritual vision: you are going to teach us, and we are going to teach you." My grandfather envisaged this marriage with his spirit and he chose the wife from the best family of the region. He analyzed family by family with his spirit to know which one was best. He chose from the family that did not like war, that did not like to fight, to gossip, to lie, and to cheat. So, he went there: "That is the family that is going to mix with my family so that good people will be born." Then we were born.
>
> We keep studying his words to see where we need to promote change to walk towards a new moment of our lives. That is why our history has brought an example. First, with the titling of our land, we united everyone to have strength to make history and defend our territory. And we managed! Our land was demarcated, with our joined forces. Then, we stopped for three years to think and investigate about everything that

there was inside our territory—good and bad things—and to think about how we could put inside us the best of all the bad moments we had faced, so that we could reflect on the changes we had to do from then on. We also thought about how we could connect to the outside world through equipment—which technology could come in so that we could communicate with the world. This was all planned. White people used to say we were thieves, that we were going to be miserable, we were going to be beggars; but we showed it to be different. We [the Ashaninka people] have learnt with nature to have everything. Our own clothes, our adornments, our chants, our myths... Everything we have are incorporations that were created by ourselves. It doesn't make sense to say we are poor. With abundant water, forest and fruit, with animals and fish, all we have is abundance. What is important is that we take care of this universe and say we are not poor: we are millionaires (Benki, August 2017).

The History of a Territory
Under Threat and Change

Ãtxoki, an Ashaninka (April 2015):

> I arrived here in 1982. I came from Peru and took part in the territorial defence. Funai also helped. It was a struggle against loggers, against slavery, against cattle rangers. In 1992, in a meeting, we decided the political objectives of the community. Even with death risks, the leaders did not give up. Pawa [their god] stated that everyone can have a space on earth and no one is the earth's owner. We are all temporary here. To live well in a space, we needed to reflect about our life project: we needed a common goal. We needed to plan it with care in order to survive.

Samuel Piyãko—an Ashaninka shaman—settled down in the Amônia River region (Acre State, Brazil) in the 1930s, after running away with his family from land invasions in other parts of the Ashaninka territory and exploitative work under a system of debt patronage, based on exchange of goods and unequal power relationships. Samuel became a leader in Amônia, gathering other families around him. He envisaged the marriage of one of his sons, Antônio, with a non-indigenous woman who lived in the region, known as Doña Piti. He understood that she was going to "look after them," as the wiracotxa ("white people") were exploiting them too much. Piti—because she spoke Portuguese and knew how to read, write, and count—became key in supporting the development of Ashaninka trade relations and in their fight against the timber patrons and the titling of their land.

In the 1980s Brazilian loggers began extracting timber from the Amônia region using heavy machinery, bringing a lot of wiracotxa to the region. The arrival of a mechanized system for timber extraction was one of the greatest shocks suffered by the population living there at the time, as it made the Ashaninka feel deeply the rapid impacts of this mode of extraction in their environment and, therefore, in their "world." More than a quarter of the land was affected. Timber extraction significantly impoverished the area's ecological diversity, impacting, for instance, the raw matter they used for kushmas ("dying clothing"), building canoes, and making drums, as well as making foods scarce. The machinery's noise scared the animals away and oil spills polluted the waters and caused a reduction in fish stocks. The loggers increased pressure on hunting and fishing stocks, further impacting food sources and resulting in widespread hunger. White men would turn up at Ashaninka piyarentsi ("traditional drinking parties"), drink spirits, and abuse the women, imposing their own musical style on the gatherings. The Ashaninka's language, clothing, and the kamarampi ("ayahuasca") ritual were discriminated against, and their practice declined. The Ashaninka began accessing white clothing and artifacts, and stopped producing many traditional pieces of craft such as bows and arrows, clothes, and adornments. Last but not least, they suffered from diseases resulting from intensified contact with outsiders, in some cases leading to deaths (Pimenta 2002: 146–149).

During this period, workers from the National Foundation for Indigenous Affairs (FUNAI) and an anthropology masters student traveled to the Amônia River and clarified its inhabitants' new land rights under the 1988 Brazilian Constitution. Chief Samuel passed away in the mid-1980s, but the struggle for territory continued. His death added to the instability caused by the struggles at the time of demarcation and led many families to abandon the area. Antônio's family was one of the few to remain. He became a key actor in the land-titling process, with Piti's support and aided by his children and family members, many of whom suffered death threats. The years of struggle cost a lot of effort, endurance, and nearly the lives of those who stood up to the state and loggers. Kampa do Rio Amônia indigenous land was, however, demarcated in 1992, comprising 87,205 hectares. When the situation in the community became more stable many families returned and Antônio was made chief. At present, there are nearly 1000[3] Ashaninka in the demarcated land, situated near the town of Marechal Thaumaturgo, on the border between Brazil and Peru. Most of them currently live in a single village called Apiwtxa.

THE LAND IS SECURE, BUT THE CLIMATE IS CHANGING

Despite their success in winning land rights and in the strategies they developed to look after the territory, which we will explore below, many threats to the mode of existence of the Ashaninka from Apiwtxa remain, such as illegal invasions. In addition, many report that the climate has become unpredictable in the area. Shamans state that something will happen, not only to the Ashaninka, but to all peoples, because Pawa ("the sun"), their major deity, was getting too hot. In the beginning Pawa lived on earth. He became hotter as he grew and had to leave otherwise he was going to incinerate the earth (Mendes 1991; Weiss 1972). Pawa is said to be unhappy with the attitude of his children and is now coming closer to earth again (Moisés in Mesquita 2012: 273).

The Ashaninka interpret weather changes in their territory by observing "bioindicators" such as the behavior of animals, plants, celestial bodies, and wind and rain. The elements of their landscape are interrelated and understanding them is crucial for survival. Collaborative research carried out with the Ashaninka of Amônia River produced the "Ashaninka calendar", an account of how they used to orient and organize their activities throughout the year based on their observations of the environment (Mendes 2002). Animals and stars indicate the beginning of seasons—that is, when certain activities, such as planting gardens, should begin; flowers and fruits would announce the right time to hunt specific animals, and birds mark the time to plant certain crops.

The Ashaninka are increasingly concerned about irregularities in these relationships, which directly affect their livelihoods as they can no longer trust in the signals they are acquainted with to carry out activities essential for their survival. They observe that fish are becoming smaller, and the river water is warmer and dirtier. They worry that the area will become a desert if the river dries up. Moisés reports that it is currently difficult to observe the celestial bodies as before because there is too much smoke in the air (Mesquita 2012: 255). Benki and Arissêmio state they are no longer able to see important stars as some of them are changing position, because the earth's axis is changing due to instability engendered by human action (Mesquita 2012: 260). Furthermore, they are concerned about the impacts of deforestation and cattle ranching around the territory and associate it with the environmental changes they are witnessing.

The explanations the Ashaninka from Apiwtxa have for these changes mingle their direct observations of predatory human action—especially deforestation and pollution—with cosmological understandings. At first, on earth there were only Ashaninka people in human form, but as they disobeyed Pawa and/or strongly expressed specific characteristics, Pawa transformed them into other-than-human beings, such as animals, plants, and other elements of their lived space like meteorological phenomena (Weiss 1972). When Pawa left the earth he transformed one of his wise daughters into kamarampi (Banisteriopsis caapi) so that she could teach and heal the Ashaninka (Mendes 2002: 185). Hence, the live elements of the Ashaninka landscape are not simply objects of consumption to satisfy their needs but "family." Their humanity is still present even though they have different bodies (Viveiros de Castro 1996). They are thus treated with collaboration and mutual respect, as the Ashaninka understand they live within a network of relationships that require constant care and attention to be sustained and well balanced.

The Ashaninka also maintain complex relationships with other invisible entities such as the enchanted beings they may interact with when drinking kamarampi or the spirit owners of animals and objects. For instance, if a man wants to hunt he is meant to ask the animal's spirit owner for permission. People are meant to consume all the animal's meat and make the most of all its elements (such as skin, feathers, and bones), as wasting can be a cause for punishment by the animal's owner and, therefore, of future deprivation (Mesquita 2012: 241–242; Beysen and Ferson 2015: 6).

Shamans (antawiari and sheripiari) are powerful individuals in Ashaninka history and are especially skilled at establishing direct communication with all those beings not present in human form. They are also amazing observers of the lives of other-than-human beings present in their landscape. In the past it is said that shamans were more powerful—the most powerful of them were the antawiari, who besides being able to communicate with other-than-human beings could perform transformations and change the course of events including meteorological phenomena. Currently, the Ashaninka recount, due to the instabilities and destruction engendered by human mistreatment of their environments their shamans are increasingly losing their power.

The Ashaninka also have a story of a great deluge, when their world nearly ended and they almost disappeared. The only Ashaninka who prepared himself for the deluge was a shaman who by drinking kamarampi knew it was going to happen. He tried to warn his Ashaninka fellows, but they were too busy drinking piyarentsi (their alcoholic beverage) and did not pay attention to his warnings. The shaman prepared himself and his family by building a boat, gathering animals and seeds, and during the rise of the waters he was able to talk to the big crab that was blocking the waters to release them (Mendes 2002: 205–208). The present-day Ashaninka are descendants of this shaman's family.

Many such Ashaninka narratives show how deities punish inattention and misbehavior, which can lead to tragedy, misfortune, or great transformation. Following such reasoning it might come as no surprise that Pawa and other spirit owners are currently punishing humanity for not looking after the earth. Guidance on how to act accordingly should be sought in spiritual practices and teachings, as we will see below.

Contemporary Thought

Moisés (May/August 2016):

> My thought is not to have a lot of money to bring the outside world here, but to strengthen our own world and to show its value to the outside. This is my grandfather's inspiration. [...] We are learning the Portuguese language to be able to explain to people about us and to understand this other world and find a way for our continuity. What is there in your world that will help us to keep our lives?

Benki and Moisés (sons of Antônio and Piti) are shamans who claim to be following their grandfather Samuel's visions. In their shamanic development they received guidance from shaman Arissêmio and, above all, from drinking kamarampi. As community leaders they have also had the opportunity to travel extensively and participate in national and international meetings concerning indigenous peoples and the environment, where they exchange knowledge with other indigenous and non-indigenous peoples. Benki and Moisés' trajectories contextualize their contemporary understanding of the challenges faced by their community and enlighten their proposals for action. Their positions also express the power of ancient Ashaninka narratives in understanding the present, in their critique of the Western world, and in their proposals for transformation.

Given their knowledge of the Portuguese language—thanks to their mother—and their experience outside their territory they have become central spokespersons for their community, acting as "diplomats", bridging the existing gaps between different worlds. Their speeches reveal a clear effort to make their reasoning and proposals for action understandable to outsiders, which recalls Arregui's exploration of the possibilities of finding common ground for discussions and possibilities of cooperation across worlds (this volume, chapter "This Mess Is a 'World'! Environmental Diplomats in the Mud of Anthropology"). I outline below some of the key aspects of their thinking that shed light on the concrete actions taken by the community.

Critique of "Science"

Benki stated (July 2015):

> A ciência sem consciência pode ser a causa de nossa destruição
> [Science without conscience may be the cause of our destruction].

Using this Portuguese wordplay Benki expresses his criticism of "science." "Conscience" here refers to the awareness of everyone's interdependency and of the need to look after life (not only human, but also other-than-human) on earth—a thought that might recall Lovelock's conception of Gaia as a living entity that is composed of other lives which are in constant interaction and make possible the existence of human beings on earth (Lovelock 2000 [1979]). As such his statement relates to a theme frequently raised in this volume, which has to do with an indigenous concern that a proper behavioural/moral code should be followed in relation to other-than-human-beings.

The kind of scientific practice Benki refers to stems from the destruction he witnessed in the Ashaninka "world" when heavy logging machinery ravaged their territory in the 1980s—that is, his experience that the power of the technology developed by science can effectively devastate the world. An Ashaninka narrative recounts that the "Inka" was an Ashaninka who had all the technological knowledge and was meant to pass it on to the Ashaninka people, but ended up trapped in the land of the wiracotxas (White men) who stole it from them (Mendes 2002: 191). The wiracotxas are said not to know how to use technology wisely and conscientiously.

Moisés, for his part, mourns the lack of respect/recognition of indigenous knowledges and points at the failure of modern Western "science" in looking after life on earth:

The world population who claim to be "civilised" do not look at the science of other peoples. Each people have their own science. This thing about burning carbon is something that my people also talk about [...] In our science we look at all that. [...] Many of our studies will be known in the coming centuries—it is knowledge that people can still not see. [...] We are certain about what we see in our traditional knowledge, and its importance is not only for the indigenous population, but is universal, for all beings in the planet and in the universe. Why do white people know that people are dying and still build things that can kill? They do not even respect their own people. If they see this is wrong, why do they do it? [...] There is something wrong with this, something that needs to be straightened. [...] I am an Ashaninka who have never been to school to study, but I know a lot of things. (Mesquita 2012: 276)

For these young shamans the real source of knowledge can be found in the spiritual world, which lays out the ethical principles that should guide relationships and how people should act on earth. Their spiritual world is their source of inspiration when it comes to making decisions and defining which paths the community should take. This world can be accessed, for instance, by following the lessons of ancient narratives or via the use of kamarampi. If developments in modern Western science have been a source of destruction, shamans claim that by (re)-connecting to the sacred and finding the right answers, they can reverse present problems. As in the flood narrative quoted above, spiritual practice can redirect action and prevent disaster.

Importance of Reciprocity and Collaboration

The shamans highlight the importance of reciprocity, another theme frequently recurring in the chapters of this volume, especially those of Permanto, Bold and Questa. As we have seen, reciprocity is a value that should not only be applied to human–human relations, but also to relations between human and other-than-human beings. According to Benki: "To exchange is an obligation of the spirit of reciprocity, a way to reconcile differences, that is, to transform. This is the Law. Today you have to learn the ancient agreement with nature, you have to learn that there must be fraternity among all beings so that Mother Earth can overcome the disease she is suffering from" (in Mesquita 2012: 306). Therefore, reciprocity among all beings is key to survival. Looking after each other (humans and non-humans) is essential. By breaking the proper relationships with other beings we can cause planetary instability. The importance of exchange is highlighted in reconciling differences and bringing about transformation. The shamans understand that the constitution of worlds is dynamic and are conscious of the importance of collaboration across them—and that this may be necessary to save the planet.

> What we are losing on Earth because of economic development is also a result of the actions of some people who became remarkable [referring to some scientists]. Here we are willing to unite everyone so that we may envisage a path for a common future. I like to speak concretely about what we have done. We have fought a peaceful fight, in which none of us died, but we have always been very strategic. We are a nation and we need to be respected as such. Thanks to our history we can reflect on the future we would like. No one from the outside needs to come here and tell us how it should be. The encounter between my mum [non-indigenous] and dad [indigenous] was a way to bring us this knowledge. My grandfather chose it to be like this. From this knowledge today, we pray for union and we need to fight prejudice. [...] Today we are here to change history, but we will not do it alone. Here we will do it differently [from modern Western science] and for the rest of the world, and people can see and copy it if they wish. Here it will be with conscience. (Benki, July 2016)

> I had a vision when drinking ayahuasca. There was a spaceship that came and took me because they wanted to show me something. It took me to the river shore and we came out and walked towards the roots of the samaúma tree [Ceiba pentandra]. "Here it is what I want to show you"—he [the spaceship being] said. He opened a triangle-like portal in the samaúma's trunk. What I saw inside was the bee's world. All the bees were working: each one of them was doing his/her part. It was like a city. Everyone who belonged to it had to do something. So, I saw all the process, and he told me: "to live well in this world, everyone must do his/her own part, exactly like this. If that does not happen, we will fail. But if everyone does his/her part, then we reach perfection". (Moisés, November 2016)

Spiritual guidance and positive thinking alone are not enough to accomplish change and transform reality. As in the narrative about the deluge and in the vision described above by Moisés, the shaman receives a warning when drinking kamarampi, but action is needed to escape trouble. Even though present-day shamans have less power to mediate between good and evil forces and intervene so that stability is maintained, Apiwtxa's shamans, along with other community leaders and members, are promoting a clear effort to change the course of things. Having their territory as a basis for action, they are seriously working on sustaining the conditions for their continuity as the Ashaninka people.

They are aware they cannot do it alone—and are attempting to lead change by example. Viveiros de Castro (Sampaio Caldeira 2018) differentiates models and examples. Models are meant to be copied. They are normative and impose ideas on people as happens in traditional models of development. Examples, by contrast, are structures of thinking based on practical experience. They may arise from not strictly following models, opening up spaces of autonomy, and may offer hints on how to act. Exemplifying with Viveiros de Castro's point that indigenous peoples can serve as examples to other peoples, the Ashaninka do not think everyone must do the same as them, but that what they do can be a source of inspiration to people on how to look after the earth.

The actions the Ashaninka from Amônia River have been taking to deal with the threats to their world since land titling—which consist in a series of adaptations in their social organization and other domains—reveal their visionary capacity to transform the situations they face, and reflect their concerted efforts not to depend on or be subject to external systems invading and threatening the continuity of their world. Their responses to changes in their environment also point to a determination to maintain the conditions in their territory for their society to flourish.

Establishing Institutions to Bridge the Gaps Between Worlds

During the struggle to obtain the land title one of the first strategies the Ashaninka adopted was to create a Cooperative, which they named "Ayõpare." The idea was to invest in marketing their products and, therefore, no longer depend on the patrons to supply them with goods, while preventing the forest from being destroyed. The working system of the Cooperative was inspired by their traditional ayõpare ("exchange system"), and consists in community members delivering what they produce and gaining credit in exchange for goods in a small Cooperative market in the village. Given this system cash was for some time not used or necessary for Amônia River community members.

The main and most successful product of the Cooperative since its creation has been Ashaninka traditional crafts. This was important for several reasons, including the income it generated for them, the fact that their production did not damage the forest, and because it meant the recovery of skills in the production of a series of items they were no longer making because of increased contact with loggers and patrons.

At first, the Cooperative assisted in commercialization by growing crops for sale in town, but soon they realized they were spending too much effort to supply an external demand for food, while at the same time impacting their own food production and increasing deforestation by opening up new gardens to cater for this trade. They decided to stop producing crops for outsiders and focused instead on their subsistence activities, based on their own rhythms and on the production of crafts to complement their own needs wherever necessary—according to each family's demand. As Moisés put it: "We got to the point of cutting the forest to plant beans and corn [to sell in town], but this did not make sense. We must look after the water, the forest, the land."

With land demarcation the Ashaninka began resorting to collective planning as a key resource to maintain themselves in their territory and protect themselves against external threats. Benki explained that they spent three years dialoguing and planning about how they were going to live in their titled land. Right after they obtained the title they founded their Association, named Apiwtxa (literally "all together"), to formally represent their interests vis-à-vis the state and civil society.

The Ashaninka from Amônia River founded their first school in the early 1990s. The idea was to promote literacy and numeracy to protect future generations from being tricked by outsiders. Since then they have been establishing a differentiated school system, in which they can teach primary education in their native tongue. Teachers come exclusively from the village and follow a specialized curriculum that balances learning about the outside with maintaining the training necessary to develop the skills for their traditional activities.

Making a Strategic Movement and Building Collective Planning Tools

The literature on the Ashaninka depicts their traditional settlements as scattered, with nuclear families establishing their houses independently (Killick 2005; Mendes 1991: 22). In the process of collective planning the Ashaninka discussed a major change and made the decision to live together, establishing the village of Apiwtxa, close to the northern limit of their land. Apiwtxa was built on two former cattle pastures that meas-ured about 40 hectares. The aims of this were to guard their territory from strangers more effectively, strengthen their collective organization, and improve natural resource management.

Right after demarcation of their land the Ashaninka developed, with the collaboration of strategic institutions, other tools that have sup-ported their planning activities, such as ethnomapping of their territory and putting in place their territorial and environmental management plan (PGTA). Ethnomapping was a way to map their territory according to their uses, values, and classifications, which helped them organize access to shared resources in a limited space. The PGTA was an instrument developed to build collective agreements on their use of territory and to express the perspectives of the community for their future. These docu-ments have facilitated the establishment of partnerships to achieve their aims and the receipt of funds to support their initiatives, both of which helped them convey their aspirations to the outside world.

Land demarcation, collective planning activities, and creation of the Association and the Apiwtxa village led to establishing a series of rules on community life based on collective agreements and the development of a shared governance system. People who did not agree with the new regulations, including families who wanted to keep on working in timber extraction and/or selling game to outsiders, had to leave the territory. Given the demand for new types of leadership (such as that required by the Association and Cooperative) and their willingness to maintain their way of life, the Ashaninka have developed a system that can respond both to the need to deal with external affairs and the complexities of life in a community with a growing population.

Reversing the Destruction of Their Land and Building Collaborations

All the changes and adaptations mentioned above served to meet one of their key concerns—namely, to reverse the destruction of their for-est so that they could live and keep alive the other-than-human beings that compose their world. The Ashaninka decided to act upon these cir-cumstances by "giving nature a hand"—as they put it—looking at both restoring and strengthening aspects of their landscape and of creating alternative sources of food so that they do not need to put too much pressure on existing resources.

They made a concerted effort to recover and restore living conditions in their territory for humans as well as for the other beings that share the habitat with them. For instance, in the 1990s they began reforesting the 40 hectares of cattle pasture where Apiwtxa village was set up by establishing agroforestry systems. They focused on planting fruit and native trees to provide them with food and timber for their own construction. These systems have furthermore been successful in attracting game back into their land and have turned it into a source of food. The Ashaninka began cultivating the plants necessary for their crafts and buildings in the gardens that surround their houses so that they do not exhaust what is in the forest. They also built fishponds so as not to put too much pressure on fish from the river. They have also succeeded in restoring the river turtle population which was on the decline.

Many of the above activities, especially at their start, were carried out on the basis of collective community action with no financial support. Notwithstanding, the Ashaninka from Apiwtxa have become increasingly skilled in raising funds and, as years passed, have been able to get support from different organizations—be they national/international, governmental/non-governmental—and even from individuals who were inspired by their example. Since the land-titling struggle the Ashaninka from Apiwtxa have been building alliances with anthropologists, governmental workers, and NGOs. These collaborations have expanded over the years and currently Apiwtxa has a wide range of partners.

Sharing Experiences and Joining Forces

The people of Apiwtxa are actively engaged in sharing their successes with other indigenous and non-indigenous communities. By promoting exchanges, courses, workshops, and collaborations the Ashaninka show how they are responding to contemporary challenges to what they want to sustain. It is also a way of making themselves understood and gaining respect, as well as to widen the scope for finding allies in their struggle for continuity. Their strategy is not to confront, but to unite and join forces.

The Ashaninka have been working with surrounding communities to share their experiences in community organization and environmental management with the ultimate objective of protecting their territory. They are specifically concerned that deforestation and other development projects may advance too much in the region and end up impacting their land. They are also working with other Ashaninka communities in Brazil and Peru, and with other indigenous communities, sharing their experience and supporting cultural revival initiatives. A very important project in this regard was the foundation of the Yorenka Atāme Center in the early 2000s in the town of Marechal Thaumaturgo, which is intended to be a training center for people from outside the indigenous land (be they indigenous or non-indigenous) to diffuse and exchange knowledge of the forest with others.

CONCLUSION

[T]here is a huge difference between being "modern" and being "contemporary". Actually, knowing how to become a contemporary, that is, of one's own time, is the most difficult thing there is. (Latour 2017: 47)

Our shared territory—which we may call "Gaia"—supports the existence of many human and other-than-human beings. Climate scientists claim human action is putting the conditions for the maintenance of many forms of life in the planet under threat. What the Ashaninka people from Amônia River show us is that it does not necessarily have to be this way.

After suffering severe threats to their existence the titling of their territory was a key achievement that enabled them to create a space of autonomy to carry out the necessary actions to prevent the destruction of their world, which depends on the maintenance of the diversity of life they interact with. It is in the Kampa do Rio Amônia indigenous land that they have been able to materialize the visions that guided their movements.

The Ashaninka strategies described here, especially since land demarcation, have to do with promoting the necessary changes to respond to contemporary challenges. Had they further engaged with loggers or chosen to invest in forest conversion in favor of cattle ranching their "wealth" would have disappeared and their "family" would have been destroyed, as it takes much more than persons to build an Ashaninka community.

Even though their actions are localized they are trying to lead by example and inspire people beyond their territory to look after the earth. Their land is a microcosm where they attempt to make it happen. As Moisés puts it: "We look at our small piece of land as if it were the planet, and we look after it as if it was our own life" (Moisés, July 2016).

Despite the uniqueness of their world they are aware it is not disconnected or isolated from other worlds that coexist with it on earth. They understand that in contemporary times the capacity to exchange and transform so that diversity can be maintained is essential for the continuity of life. Cooperation among all beings is important so that our shared living space will continue to sustain us:

> Our life is a dream. The dream is a continuity of our life. We also dream awake with the continuity of our life. What is happening here is the continuation of our grandfather's dream. Here we are [his grandchildren] accomplishing what he thought: guaranteeing the continuity of our people and constructing the best path to all of us. (Moisés, April 2016).

NOTES

1. There are four shamans key to this chapter: (1) Samuel: the shaman-leader founded the current settlement in Amônia River; (2) Arissêmio: an elder shaman married to one of Samuel's daughters; and (3 and 4) Benki and Moisés: Samuel's grandsons and present leaders in Amônia River. I have personally met Arissêmio, Benki, and Moisés. In this chapter, I am highlighting the shaman's voices not only due to their central role in Ashaninka society, but also because their narratives stand out in the debates relevant to this chapter. That does not diminish the importance of the role of the whole community and other key leaders in resisting and performing the transformations central to their struggle.
2. In Peru the estimate is 141,183 people (PERU, 2017, http://bdpi.cultura.gob.pe/pueblo/ashaninka) and in Brazil 1645 (Siasi/Sesai, 2014, https://pib.socioambiental.org/pt/Povo:Ashaninka).
3. In 2014 the official number was 940 (Siasi/Sesai, 2014, https://terrasindigenas.org.br/es/terras-indigenas/3716#demografia).

REFERENCES

Beysen, Peter, and Sonja Ferson. 2015. *Ashaninka: O Poder Da Beleza*. Rio de Janeiro: Museu do Índio.

Escobar, Arturo. 2017. "Sustaining the Pluriverse: The Political Ontology of Territorial Struggles in Latin America." In *The Anthropology of Sustainability*, 237–56. New York: Palgrave Macmillan. https://doi.org/10.1057/978-1-137-56636-2_14.

Killick, Evan. 2005. "Living Apart: Separation and Sociality Amongst the Ashéninka of Peruvian Amazonia." PhD thesis, The London School of Economics and Political Science.

Latour, Bruno. 2017. "Anthropology at the Time of the Anthropocene: A Personal View of What Is to Be Studied." In *The Anthropology of Sustainability*, 35–49. New York: Palgrave Macmillan. https://doi.org/10. 1057/978-1-137-56636-2_2.

Lovelock, James. 2000. *Gaia: A New Look at Life on Earth*. New York: Oxford University Press. https://doi.org/10.1016/0004-6981(80)90149-3.

Mendes, M. K. 1991. "Etnografia Preliminar Dos Ashaninka Da Amazônia Brasileira." UNICAMP.

Mendes, Margarete K. 2002. "O Clima, o Tempo e Os Calendários Ashaninkas." In *Enciclopédia Da Floresta - o Alto Juruá: Práticas e Conhecimentos Das Populações*, edited by M. C. da Cunha and M. W. B. de Almeida, 179–220. São Paulo: Companhia das Letras.

Mesquita, Erika. 2012. "Ver de Perto Pra Contar de Certo. As Mudanças Climáticas Sob Os Olhares Dos Moradores Da Floresta Do Alto Juruá." UNICAMP, Campinas University.

Pimenta, J. 2002. "'Índio Não é Todo Igual': A Construção Ashaninka Da História e Da Política Interétnica." Universidade de Brasília.

Piyako, F., B. Piyako, W. Piyako, M. N. Samaniego Pascual, J. Koshipirinke, D. S. Salisbury, and C. S. Comandulli. 2015. "Ashaninka People of the Amazon Unite to Face Climate Change and Extractive Development." https://blog.richmond.edu/dsalisbury/files/2011/08/AshaninkaPoster20151122A0English.pdf.

Ricardo, Beto, and Fany Ricardo, eds. 2017. *Povos Indígenas No Brasil 2011/2016*. Edited by Beto Ricardo and Fany Ricardo. São Paulo: Instituto Socioambiental - ISA.

Sampaio Caldeira, Ana Paula. 2018. "Sobre Modelos e Exemplos." *Varia Historia* 34: 9–12.

Viveiros de Castro, Eduardo. 1996. "Images of Nature and Society in Amazonia." *Annual Review of Anthropology* 87: 179–200.

Weiss, Gerald. 1972. "Campa Cosmology." *Ethnology* 11 (2): 157–172.

Relational Ecologists Facing "the End of *a* World": Inner Transition, Ecospirituality, and the Ontological Debate

Jean Chamel

The ecologists[1] who show concern for the future of the planet, worrying about climate change and the potential collapse of our complex societies, are often depicted as prophets of doom predicting the end of the world and ridiculed for their apocalyptic views. The derogatory titles of several recent publications, such as "Apocalypse Fanaticism" (*Le Fanatisme de l'apocalypse*, Bruckner 2011), "Preachers of the Apocalypse" (*Les Prêcheurs de l'apocalypse*, Kervasdoué 2007), or even "The Apocalypse Isn't Coming Tomorrow" (*L'Apocalypse n'est pas pour demain*, Tertrais 2011) illustrate this controversy.

The study of the narratives and practices of an informal network of ecologists living in France and Switzerland gives a more complex picture. These highly educated ecologists acting as the intellectuals of French-speaking

J. Chamel (✉)
Département Homme et Environnement, Muséum National d'Histoire Naturelle, Paris, France

Institut d'Histoire et Anthropologie des Religions, Université de Lausanne, Lausanne, Switzerland

© The Author(s) 2019   161
R. Bold (ed.), *Indigenous Perceptions of the End of the World*,
Palgrave Studies in Anthropology of Sustainability,
https://doi.org/10.1007/978-3-030-13860-8_8

environmentalism (or at least its "enlightened catastrophist" element) are mainly concerned with the future of "thermo-industrial civilization" and expecting its collapse. But they deny that they are waiting for the end of *the* world. They rather expect the end of *a* world and the start of a new one.

These ecologists are linked to the Transition Towns movement that emerged in Totnes (Devon) 10 years ago and, as we will see, are advocating a transition toward a sustainable society, a metamorphosis that requires radical political change, but also personal transformation. Linking the fate of the "earth system" Gaia, somehow viewed as a living entity, to "Inner Transition," they draw correspondences throughout the "web of life" and commonly refer to the movements of deep ecology and ecopsychology. To better embrace the complexity of their worldview this chapter will examine how the latter can be related to Western esotericism as Antoine Faivre defines it. We will then enter the "ontological debate" to see where the representations of these ecologists would best fit in the ontological grid developed by Philippe Descola.

An Informal Network of "Relational Ecology"

This informal network links a hundred actors, identified by snowball sampling, whose discourse production (books, articles, blogs, magazines, conferences, forums, etc.) is the common feature. These actors belong to various interconnected networks, including the *Institut Momentum*, *Terr'Eveille*, and *Chrétiens Unis pour la Terre*, and a more informal network of ecopsychology in French-speaking Switzerland.

Many of these ecologists express strong affinities with the Transition Network (and its spiritual component, the "Inner Transition") and the *Colibris* (Hummingbirds) movements, respectively inspired by the British permaculturist Rob Hopkins and the French Algerian agroecologist Pierre Rabhi. These actors are mostly between 30 and 60 years old and two-thirds of them are men. They come from the upper middle class and have strong educational and cultural capital (almost all have a Masters or equivalent, a third hold a Ph.D.). They are journalists for national newspapers, former ministers, university professors, engineers, climate specialists (some contributing to the IPCC reports), but also independent journalists, consultants, bloggers, book writers, and speakers. They read and write profusely, often express themselves in public, and are the intellectuals of a specific ecology, which combines an *apocalyptic* perspective, the expectation of "*collapse*," and a *holistic* "*eco-spiritual*" view of the world, according to which "everything is connected." For these reasons I have decided to call them a *relational ecology* network, a phrasing that resonates with the "Amerindian landscapes" that for Bold "all consist of interdependent and mutually sustaining beings" (Bold, chapter, "Reciprocal Subsistence").

My research focuses mainly on their discourse, based on data collected through more than 50 semi-structured interviews, as well as the analysis of a large corpus of public writing and videos from those who enjoy some visibility among the larger French-speaking environmentalist milieu.[2] I also conducted participant observation in various settings such as conferences, workshops, or festivals. In addition, I carried out ethnographic fieldwork in the Findhorn Foundation in Scotland and the *Hameau des Buis* in Ardèche, eco-villages where several of these relational ecologists had lived or stayed, as well as the eco-villages of *Torri Superiore* in Liguria and *Huehuecoyotl* near Mexico City. I also participated in Belgium in a deep ecology workshop widely popular among the network, "The Work that Reconnects," inspired by the work of the US eco-activist Joanna Macy.[3]

THE END OF A WORLD AND THE TRANSITION

The members of this relational ecology network share a common statement: "we are heading straight for disaster". Such catastrophism is assumed, and many define themselves as "enlightened catastrophists," a term coined by philosopher Jean-Pierre Dupuy (2002). They are trying to think through the expected "collapse," with many references to Jared Diamond's book *Collapse* (2005), but also *The Collapse of Complex Societies* (Tainter 1990), *The Collapse of the Western Civilization* (Oreskes and Conway 2014), and *Comment tout peut s'effondrer* ("How Everything Can Collapse") (Servigne and Stevens 2015), whose authors forged the term "collapsology" to designate what they consider a new field of study.

Most of these catastrophists, however, or indeed maybe all of them don't see this collapse, be it triggered by climate change, "peak oil," economic crisis, population growth, or actually a combination of all these factors, as the "end of the world," but rather as the end of *a* world. There is an *after*, a post-collapse perspective, or the hope that the collapse can be prevented by changing the world now. All of this can be related to apocalypticism and variations on pre- and post-millenarianism, historical notions that are mistakenly associated with the idea of the "end of the world" (Hartog 2014), while they all refer to the perspective of *passage* from one world to another (Chamel 2016). Therefore, since it is not the end of the world that is at stake it is then necessary to prepare for the coming world, and preferably to realize a *transition* toward it. Such a transition is thought by most to be a better prospect than the post-apocalyptic world that some "survivalists" envision.[4]

In this sense the Transition Towns movement that started in Totnes in the first decade of the twenty first century is a clear reference point for a large part of the network, founded on the idea of starting to live a "post-oil" life. This movement primarily promotes the relocation of economic flows to create sustainable and "resilient" communities that would be able to face (or in the most optimistic scenario, to prevent) the collapse of "thermo-industrial civilization," but many actors in the Transition Towns movement also promote an "Inner Transition" that would complement what they call the "Outer Transition." Citing Ghandi's famous quote "Be the change you want to see in the world," the webpage of an Inner Transition group in Romandy explains that:

> [...] if the world shapes us, we also shape the world in our own image. Consequently, if we want to achieve real changes in our individual and collective behaviour, it is necessary to act not only on society but also on ourselves. The Inner Transition is this work of consciousness and transformation.[5]

The existence of this inner perspective indicates that an understanding of these ecologists cannot be limited to their expectation of the collapse. Such a personal transformation could open up other levels of understanding and therefore requires further exploration, with the necessity to better understand how the actors perceive and relate to the world, to embrace their own cosmology.

FROM SYSTEMIC APPROACHES OF COLLAPSE TO HOLISTIC–MONISTIC VIEWS AND ECOSPIRITUALITY

The actors generally refer to scientific, or at least para-scientific, theories with a particular insistence on *systemic* approaches, as they tend to understand everything in terms of "systems": ecosystems, of course, to understand how "nature" works and is threatened by human activities, but also the all-encompassing "Earth System." They often speak of the "thermo-industrial system" to designate capitalist/industrialized societies, and they also tend to denounce the "system" with the same word, as can be read on the placards of demonstrations that accompany UN conferences on climate change: "Change the system, not the climate." But the notion of system goes beyond the use of such expressions. These ecologists frequently speak of "complexity" as conceptualized by the sociologist Edgar Morin (2008), who is a strong reference point for many of them. They also have good knowledge of system theories, citing authors such as Joël de Rosnay (2014) or Jean-Louis Vullierme (1989).

Their interest in systemic approaches is strongly related to their perception of the expected forthcoming collapse, especially by the references they often make to the report to the Club of Rome *Limits to Growth* (Meadows et al. 1972). This report is the description of one of the first computer models, *World 3*, that tried to describe the future of the world by considering the cybernetic interactions of diverse factors such as world population, pollution, available food, and industrial production. Such a model is a direct legacy of the cybernetic and systemic theories elaborated in previous decades.

But many actors in this environmentalist intelligentsia go further than referring to systemic approaches. They also speak a lot about *holism*.[6] For them it is more or less the same thing: they want to emphasize that the whole is more than the sum of its parts. They like to say that they have developed an encompassing global vision, and place a strong emphasis on denouncing "reductionism," which they strictly oppose to holism.

For them, criticizing reductionism means criticizing Descartes, with particular emphasis on his famous quote about humans "becoming masters and possessors of nature." Through holism they reject the separation between humans and non-humans, and therefore they discard the very concept of "nature," often referring to the work of Bruno Latour (1993) and Philippe Descola (2013):

> You know Latour's book *We Have Never Been Modern*, this reading has been an important step of my evolution, the fake narrative of the Nature/Culture divide, it really struck me because I had never questioned that before. (Environmental Journalist, Paris)

> When I read Descola, I really understood that strict dualism was really driving us straight to disaster. [...] So to overcome the Nature/Culture divide, we must get rid of this Judeo-Christian distinction between nature

and man that dominates it. (Engineer who became an independent researcher and permaculturist, Paris)

As an alternative to such dichotomies they promote the idea of inter-connection between all living beings, and many like using the expression "web of life," which is the title of a book by Frijtof Capra (1997). The actors demonstrate affinities with two schools of ecological thought: *deep ecology* and *ecopsychology*. Both promote the idea of an "extended self" or "enlarged self" that encompasses the entire universe. In such a perspec-tive the concept of ego is considered too limited: the self depends too heavily on interactions with other beings to be seen as an independent entity. Such views lead to a form of spiritual ecology, or *ecospirituality*, centered on the view that everything is connected to everything else. It is in this way that these ecologists define spirituality: not in terms of a spe-cial relation with God, but as all these networks of relationships that link beings, human and non-human.

Ecospirituality can be interpreted as a *monism*: for many actors in this network of relational ecology the connection between all beings goes beyond physical interrelation: they also defend the idea that there is a unity of all things, with "something" that circulates between them, be it called "energies," a "flow," an "intelligence," or "vibrations." This is the fundamental idea of the ecopsychology school, founded by Theodore Roszak, who was inspired by Carl Gustav Jung and his concept of the collective unconscious. For Roszak there is an "ecological unconscious": an *anima mundi*, a soul of the world. Our individual health and destiny are therefore intimately linked to those of the Earth. This leads back to the "Inner Transition," which is deeply linked with the Transition Towns movement and with the practical ecopsychology developed by Joanna Macy through her "Work that Reconnects" method. For the "inner transitioners," Outer Transition cannot happen if there is not an Inner Transition, a personal, spiritual change that makes full sense in the monis-tic approach of ecopsychology: "Restoring the Earth" occurs by "Healing the Mind," claims the title of a book edited by Roszak et al. (1995).

Finally, these elements can be linked to the Gaia theory (or hypothe-sis). In its original formulation, proposed by James Lovelock and Lynn Margulis, it derives from system theories on cybernetic feedback loops, and suggests that living organisms interact with each other and with their inorganic surroundings on earth to form a self-regulating, complex sys-tem that helps to maintain the conditions for life on the planet (Lovelock and Margulis 1974a, b). Later advocates of the theory read into it another level of understanding, considering Gaia as a living being. Gaia is not seen as a deity to be worshipped, but can be understood as the entity where the "web of life" is deployed, or assimilated to the *anima mundi* mentioned before.[7]

As we have seen, the actors seize upon systemic approaches to make sense of the expected collapse and recognize them as well in the Gaia theory, but this systematic thinking is also a gateway to holism. These ecologists share a holistic and monistic worldview that refers to the deep ecology and ecopsychology movements, and an ecospirituality that intertwines personal and planetary transformation, inner and outer transition. These outcomes prove that there is much more to be understood about the catastrophism of the actors than a mere waiting for the "end of the world." The various references listed above suggest that fruitful comparisons could be drawn with Western esotericism.

Western Esotericism, Universal Correspondences, and the Ontology of Analogy

The historian Antoine Faivre has proposed a morphological definition of modern Western esotericism to identify what he calls an "esoteric form of thought" in different temporal contexts. This definition has four essential characteristics:

1. *The idea of universal correspondence*: "there are non-causal correspondences between all levels of reality in the universe, which is a kind of mirror theatre browsed and animated by invisible forces," including relations between the sky (macrocosm) and man (microcosm).
2. *The idea of Living Nature*: "the cosmos is not only a set of correspondences. Ridden by invisible but active forces, the whole of Nature, considered as a living organism, as a person, has a history, linked to those of humans and of the divine world." Faivre also adds the idea of a "Suffering Nature" awaiting its deliverance.
3. *The role of mediation and imagination*: "rituals, symbols with many meanings (a mandala, a tarot game, a biblical verse, etc.), intermediate spirits (such as angels) are mediations capable of securing passages between the various levels of reality. The 'active' imagination (or 'creative' or 'magic' imagination, a specific—but generally dormant—faculty of the human mind) exercised over these mediations makes them a tool of knowledge (tool of 'gnosis'), or even of 'magic' action, on reality."

4. *The experience of transmutation*: "it is the transmutation of oneself, which can be a 'second birth,' and in corollary that of a part of Nature (in many alchemical texts, for example)." (Faivre 2007: 15–16)

Among these four characteristics the *idea of "Living Nature"* can most immediately be related to the Gaia hypothesis and the organic conception of Gaia as a "Living Earth" since the actors in question generally agree to consider the Earth–Gaia a "living planet." This similarity has been emphasized by the Germanist Aurélie Choné in her definition of ecospirituality (Choné 2016: 63) as well as by the political scientist David Bisson in a text on spiritual ecology. Bisson does not separate the idea of Living Nature from that of *universal correspondence*. According to him both "represent nature as a theatre of mirrors, a mosaic of symbols that ultimately reveals the harmonious unity of man and the universe" (Bisson 2013: 165–166).

In fact, the organicism of the relational ecologists applies beyond Gaia, at various scales. All of perceptible reality and beyond is then organized in an organic way, in different "subsystems," by a system of correspondences:

> I think that the Earth is a multicellular organism that includes human beings and all other forms of life, and I think that the Earth has its identity too, its personality. There are probably other planets that are alive, maybe another system beyond. I think the planet is alive, it has its personality, its identity, and I'm part of it, I am a subsystem of the Gaia system. I have myself a subsystem of billions of living cells, which I thank. There are different levels, we go up a notch each time, but everything is connected. (Chemist who started an ecological community near Geneva)

Each "level" is made of various "subsystems" and is itself one of the many "subsystems" of a higher system. Thus different "cellular" and "living" systems fit together like Russian dolls in perceptions of the world as a "fractal" set (see, Bold, chapter, "Pesticides and the Fractal Landscape" section for another account of a "fractal landscape"). Humans are composed of many cells, and constitute the cells of a larger body: humanity. And in turn all living beings are the cells of an even wider living organism, the Earth. The Earth itself, like other "living planets," could be part of an "other system beyond." The existence of a single principle of organization repeating itself at all scales expresses the monistic idea of unity beyond the perspective of interconnection.

It appears then that the cosmology of these ecologists includes a *principle of universal correspondence*.

The *experience of transmutation*, both of oneself and of nature, is also a notion that emerges clearly from discourse and practice. The idea of the transformation of the world is at the heart of the Transition Network, and the Inner Transition complements it with the transformation of self. The ideas of "mutation of consciousness" and "metamorphosis" are commonplace in these discourses, and transmutation is even experienced in practices such as the composting of food wastes, "a form of alchemy" for an actor, or the use of composting toilets (Chamel 2017):

> What some people are experimenting with composts, it is very, very powerful. They experience the mystery of a disgusting thing that becomes a wonderful thing, and it's bluffing, it's almost a metaphysical experience. We are all compost. (Lawyer, head of a small eco-Christian association, Paris)

Finally, the *role of mediation and imagination* is more difficult to identify, though not absent from discourse and practice. Imagination and creativity are indeed qualities commonly promoted within the network, and more particularly by those who claim to be "cultural creatives." The exercises and rituals of "(re)connexion," especially those of the "Work that Reconnects," adopted by the actors, highlight even more the function of mediation within the practices. Some of them, speaking of "Goethean sciences," also strongly defend the role of intuition and creative imagination, as a "tool of knowledge."

The examination of the discourse of the actors through the prism of Faivre's four characteristics therefore presents strong affinities between these discourses and several elements that characterize Western esotericism for the French historian. The point here is not to discuss whether ecospirituality is, or is not, an "esoteric form of thought." Actually Faivre's definition has drawn criticism from other specialists on esotericism such as Kocku von Stuckrad (2005) and is not definitive. It should however be considered as a "provisional heuristic intended to revive methodological reflection" (Faivre 2007: 22) that can bring a deeper understanding of relational ecology. The Gaia theory associated with the idea of a living planet reaches a more complex dimension when it is considered with organicist discourses associated with the idea of universal correspondence.[8] These aspects, as well as the experience of transmutation and, to a lesser extent, the role of imagination and mediation, can be understood as making sense *together*, rather than being scattered, isolated with only a small interpretive range.

In particular, Faivre's definition is helpful in pointing out this dimension of correspondences, which is generally invisible in the literature on environmentalists. We will therefore examine whether such an idea of universal correspondences can be related to the ontology of analogy that Philippe Descola ascribes to many cultural contexts, including pre-naturalistic Europe.

THE ONTOLOGICAL DEBATE

After having sought to better understand ecospirituality by relating it to Western esotericism it could be helpful to analyze it in ontological terms, using in particular the modes of identification proposed by Philippe Descola. In *Beyond Nature and Culture* (2013) Descola identifies four "ontologies" or "modes of identification" to classify the various "compositions of the world" made by human societies at all times and in all places. His grid of analysis is based on the separation, considered universally shared, between what he calls "interiority" and "physicality" (Descola 2005: 210–211). From this distinction, four combinations are possible according to the perception of continuities or discontinuities in terms of interiority and physicality, in any relation to otherness:

> The recognized formulae for expressing the combination of interiority and physicality are very limited. Faced with some other entity, human or non-human, I can assume either that it possesses elements of physicality and interiority identical to my own, that both its interiority and its physicality are distinct from mine, that we have similar internalities and different physicalities, or, finally, that our interiorities are different and our physicalities are analogous. I shall call the first combination "totemism," the second "analogism," the third "animism," and the fourth "naturalism" (Fig. 1). These principles of identification define four major types of ontology, that is to say, systems of the properties of existing beings; and these serve as a point of reference for contrasting forms of cosmologies, models of social links, and theories of identity and alterity. (Descola 2013: 121)

Descola identifies naturalism with the "modern ontology" and attributes the mode of identification that is structurally symmetrical,

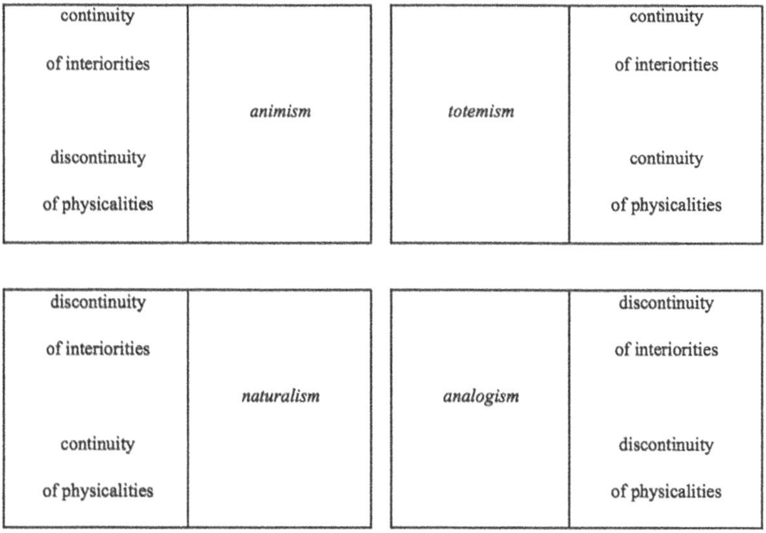

continuity of interiorities	*animism*	*totemism*	continuity of interiorities
discontinuity of physicalities			continuity of physicalities
discontinuity of interiorities	*naturalism*	*analogism*	discontinuity of interiorities
continuity of physicalities			discontinuity of physicalities

Fig. 1 The four ontologies described by Philippe Descola (2005: 221, *my translation*)

animism, to the Amazonian "collectives" he knows well. The other two ontologies of the quadrant, totemism and analogism, are respectively associated with the aboriginal collectives of Australia and with a patchwork of civilizations including Europe of the Renaissance, ancient China, Brahmanic India, West Africa, the Andes, and pre-Hispanic Mesoamerica.

For Descola, modern Western societies are deeply rooted in a naturalistic ontology characterized by linking humans and non-humans in a material continuity while attributing to them different interiorities (the cultural aptitude of humans is denied to non-humans). He concedes, however, that there are "variations" within naturalism, and their very existence, as well as

> their increasing numbers over the past decades in themselves suggest that the naturalist schema can no longer be taken for granted (which is why a book such as this one has now become possible) and that a phase of ontological recomposition may be beginning, the results of which are as yet unpredictable. (Descola 2013: 198)

Could we then envisage that the discourses of the relational ecology network herald some of these variations? If so, what would be the other mode of identification toward which this system of representations would lead? Animism is then the first that comes to mind, as the actors use it to express their own distance from naturalistic ontology, such as this actor, who defines so his relation to transcendence and the non-human: "… in fact I am almost an animist, I could almost see a god in every living being."

In Descola's structural schema, however, animism is in symmetrical (structural) opposition to naturalism, and their—theoretical—total contrast seems to leave little room for a gradual displacement of naturalism toward animism sensu Descola. But these structural categories, as "good to think with" as they might be, are considered too restrictive by other anthropologists. For instance, Yates-Doerr and Mol find Descola's naturalism too uniform and rigid to account for "Western" practices, which are, according to them, "complex juxtapositions of different modes of ordering—containing contrasting and overlapping repertoires" (Yates-Doerr and Mol 2012: 58). Their critique does not suggest another categorization to enhance Descola's work but rather emphasizes the difficulty of using such categories when examining empirical contexts. *A contrario*, Marshall Sahlins suggests reorganizing Descola's grid by considering animism, totemism, and analogism as three variants of one single animistic ontology opposed to naturalism. He therefore proposes to rename the three, respectively, "communal animism," "segmentary animism," and "hierarchical animism":

> Rather than radically distinct ontologies, [sic] here are so many different organizations of the same animic principles. Classical animism is a communal form, in the sense that all human individuals share essentially the same kinds of relationships to all non-human persons. Totemism is segmentary animism, in the sense that different non-human persons, as species-beings, are substantively identified with different human collectives, such as lineages and clans. (Apologies to Marx for this adaptation of "species-being.") Analogism is hierarchical animism, in the sense that the differentiated plenitude of what there is is encompassed in the being of cosmocratic god-persons and manifest as so many instantiations of the anthropomorphic deity.[9] (Sahlins 2014: 281–282)

The animism claimed by some of the relational ecologists would therefore rather be a "hierarchical animism" specific to analogism since their holistic and organicist views seem to correspond more to the analogist ontology, which is defined by Descola as

> a mode of identification that divides up the whole collection of existing beings into a multiplicity of essences, forms, and substances separated by small distinctions and sometimes arranged on a graduated scale so that it becomes possible to recompose the system of initial contrasts into a dense network of analogies that link together the intrinsic properties of the entities that are distinguished in it. (Descola 2013: 201)

Analogism therefore frames the world as a complex mesh of analogies necessary to connect entities distinct in essence (the discontinuity of interiorities and of physicalities). The organicism of ecospirituality, and the idea it shares correspondences between the different scales of the universe, from the microcosm to the macrocosm, are typical analogist elements that suggest that ecospirituality should rather be associated to the analogist mode of identification. These elements were central components of the pre-modern European *Weltanschauungen* (sometimes cited by the actors), those forerunners of the naturalistic "Great Divide" between nature and culture, which Descola associates with the ontology of analogy. While this category is adopted by some scholars, who find it enhances their understanding of the societies they study (Albert 2009; Tan 2012), others find it problematic for their fieldwork. For instance, Stephan Feuchtwang considers analogism "too ontological, too rigid, too ahistorical" (Feuchtwang 2014) and Descola can only reply to his objection by repeating that this category, however broad it might be, still makes sense in contrast with the other modes of identification (Descola 2014b). But the point here is not to review case studies to discuss the limits of a category that seems effectively overstretched (and therefore lose consistency), nor to take it for granted in a way that has little heuristic value, but to take it as it is to deepen our understanding of relational ecology, and then to consider how it could be re-framed to strengthen Descola's scheme.

In a book of interviews with Pierre Charbonnier, Descola tends to prophesize the emergence of a "new analogism," while he was previously more skeptical about the possibility of a surpassing naturalism[10]:

> It may then be asked whether a new constitution can truly appear and stabilize on the basis of a mode of identification whose evolution is necessarily slower and uncertain. I think this is not impossible, and it can take the

form of a passage to the limit of the possibilities of naturalism. One can conceive of a system in which interiority would no longer be conceived of as a block, as the distinctive property of a being absolutely different from others. On the contrary an evanescent and pluralistic interiority would be distributed in the whole of things, with a background of universal physicality, in a manner quite similar to what is observed in analogism. This new analogism would thus be characterized by the diffraction of the value formerly accorded to human interiority in a wider and more open range of non-human beings, but in which the scientific method based on the universality of the laws of matter would still be valid. (Descola 2014a: 302, *my translation*).

For Descola, "it seems very unlikely that we will return to animism or totemism, which are now profoundly incompatible with irreversible elements of our society." He therefore sees the development of a "new analogism" as the most probable and the most desirable "insofar as it is the most able to integrate the non-human in our socio-political constitution" (Descola 2014a: 303, *my translation*). This "new analogism" would combine at the same time the analogist elements that had hitherto been associated with esotericism, and the naturalistic validation of "the scientific method based on the universality of the laws of matter" in the words of Descola. Wouter J. Hanegraaff's understanding of the New Age—to which ecospirituality can be related—as a "secularized" esotericism, that has integrated the scientific principle of causality since occultism (Hanegraaff 1998: 441–442), supports Descola's insight. But would such a transformation make modern societies return from the "naturalism" box to the "analogism" box?

If we come back to the operation of the "Great Divide," which very gradually transformed the mode of identification of European societies from analogist to naturalist, this transition in Descolian terms resulted in a change in the apprehension of human and non-human physicalities: beforehand perceived as different and discontinuous and afterwards understood as belonging to the same continuum of physical realities. At the same time there remained a dissimilarity of interiorities between humans, beings of culture, and non-humans, conceived as integrated with "nature" and devoid of interiority or at least of an interiority comparable with that of the human spirit.

Now, if a "new analogism" is emerging, it is characterized by the persistence of the scientific method, hence the postulate of a continuity of physicalities, of a "universal physicality" (under the auspices of the universal laws of physics), and in parallel "an evanescent and pluralistic interiority would be distributed in all things," to cite Descola once again. In other words, we would go step by step toward a continuum of interiorities. Resemblance of physicalities and resemblance of interiorities: this "new analogism" should therefore very logically find its place in the "totemism" box of the ontological grid of Descola!

If the above interpretation is correct, Descola's ontology combining the continuity of both interiorities and physicalities ("totemism") deserves to be renamed to include this emerging "composition of the worlds," which is certainly very different from Australian totemic ontology. Reciprocally, this "new analogism" would show that the mode of identification mixing the discontinuity of interiorities and physicalities does not have the monopoly of analogical thinking to relate beings through correspondences. There would thus be at least two analogisms, one hierarchical—the "Great Chain of Being" described by Arthur Lovejoy (1936) and non-causal—corresponding to Descola's original analogist ontology, and another pretending to scientific universality (without necessarily confining itself to "dominant" science) and to the equality of beings in interdependence within the "web of life" (the biocentric egalitarianism of deep ecology), which would share with totemism (a segmental animism, as Sahlins suggests?) the same continuist ontology.

Such a perspective is only sketched out here and should merit further study. It could fit with the "logic of the same" identified by the anthropologist Vassos Argyrou in *The Logic of Environmentalism* (2005). Argyrou proposes to go beyond the "phenomenological" perception that sees ecology as a break with the modernist paradigm, to show that the former does in fact push a similar logic, that of the "same":

> [...] the apologists of the modernist paradigm do not fully understand the logic of the culture they champion. Environmentalists do not deny Humanism. They are far too modernist to tolerate any sort of human division. They are extending, broadening, widening, universalising more, totalising at an even grander scale. They are striving to efface the last "Great Divide" of the modernist paradigm—that between humanity and nature—to unify the world at the most inclusive level possible, stretching in the process Humanism beyond itself so that it no longer recognises itself. Environmentalists do nothing more than to take the modernist logic of the Same to its logical and onto-logical extreme. With environmentalism, modernist culture becomes fully articulated and further entrenched. (Argyrou 2005: 117–118)

In ontological terms the "logic of the same" would thus have allowed us to pass initially from "a mode of identification that divides up the whole collection of existing beings into a multiplicity of essences, forms, and substances separated by small distinctions" that are recomposed "into a dense network of analogies" (Descola 2013: 201), to the naturalism of the "modernist paradigm" that establishes a continuity between physicalities while preserving the dissimilarity of interiorities. The ecologists—and the actors of the relational ecology network in the forefront—would then pursue this logic of the same by seeking to establish a continuity of interiorities without challenging the continuity of physicalities.

Conclusion

The catastrophism of the actors of this informal network of relational ecology has thus led us beyond the understanding of their views as ecological apocalyptic or millenarianism. Since they do not see the collapse as the end of the world but the end of *a* world they engage in a transition process, which is not only practical and material, but also internal and spiritual, comprising an "Inner Transition." While the collapse is envisioned through a systemic approach, these ecologists tend to develop a holistic and monistic worldview. Their way of thinking can then be labeled "ecospirituality," a spiritual ecology that embraces the movements of deep ecology and ecopsychology, as well as the Gaia theory.

Ecospirituality, as we have seen, can be related to Western esotericism through an examination of the characteristics proposed by Antoine Faivre. The comparison is of heuristic interest to access a deeper understanding of some aspects of the representations of these ecologists: their organicism (the Earth–Gaia as a living planet), their interest in the transmutation processes ("Inner" and "Outer Transition"), but also the importance they give to correspondences. This last element led then to the "ontological debate," to see with which "mode of identification" ecospirituality would best fit. This discussion finally showed that the ontological grid proposed by Philippe Descola should be re-framed to better take account of a "new analogism" expressed in particular by these ecologists.

All these elements might be of interest to feed an emerging Anthropology of Sustainability (Brightman and Lewis (eds.) 2017).

As Moore called in this volume for "an understanding of interrelated social and environmental change that accounts for the complex temporalities and spatialities of interdependent bodies, agencies and environments, from the microbial to the planetary scale" (Moore 2017), relational ecologists offer in their discourse patterns for understanding this complexity that this field of anthropology should not ignore.

NOTES

1. I translate the French word "écologiste" into "ecologist" and not, as is usually done, into "environmentalist" because environmentalism implies an anthropocentric stance where humans are central and surrounded by their "environment." *A contrario* the ecologists considered in this chapter see the world as a "web of life", a relational ecology in which humans are part of "nature" without any privilege.

2. The discourse of these intellectuals is well known within this milieu but has a much smaller impact outside due to its incompatibility with the dominant narrative of growth, progress, and modernity. This inconsistency incites these ecologists to keep their distance from the "mainstream" and get closer to countercultural currents.

3. This research was part of a Ph.D. project defended in 2018 at the University of Lausanne (Chamel 2018).

4. The "end of a world" pattern can be easily associated with millenarianism, a term that applies to "all movements and organizations that hold as a central belief the imminent arrival of a divinely inspired and this worldly society, whether a religious golden age, messianic kingdom, return to paradise, or egalitarian order. Such movements can take on an active or passive, violent or peaceful, even revolutionary role. They are found the world over and throughout recorded history." This broad definition of the iResearchNet dictionary of sociology of religions (http://sociology. iresearchnet.com/sociology-of-religion/millenarianism, June 2018) is rather vague because this term refers to such a large range of political and religion movements over time and space that its heuristic usefulness is not demonstrated. In general, historical and ethnographic descriptions avoid engaging in providing a clear definition of millenarianism. In the European Christian context, where the term was coined, *The Pursuit of the Millennium* of Norman Cohn (1957) is the most obvious reference. *The Trumpet Shall Sound* of Peter Worsley (1957) is another classic piece about cargo cult in Melanesia, with a biblical reference in the title linking the contexts.

5. http://www.ecoattitude.org/accueil/transitioninterieure. Accessed October 17, 2015.

6. Holism can be understood as the tendency to embrace the whole in a single move, while systemic approaches instead consider all the interrelations of the parts to apprehend the whole that they constitute together.
7. Bruno Latour developed an interesting reflection around this issue, though not based on ethnographic material (Latour 2017). It seems that Lovelock never presented Gaia as a "God of Totality" and always denied any divine interpretation of Gaia. He however often played with the religious lexical field entitling, for instance, a chapter of one of his books "God and Gaia" (Lovelock 1988). Lovelock, although avoiding presenting Gaia as a living organism, also allowed the French translation of his book *Gaia: A New Look at Life on Earth* to be entitled *La Terre est un être vivant* ("Earth Is a Living Being") (Lovelock 1986).
8. A conception that exceeds by far Lovelock's understanding of Gaia.
9. In another way, Nurit Bird-David proposed before Descola published his work on the modes of identification, to understand animism as a "relational epistemology" (Bird-David 1999) instead of an ontology.
10. "To effect a deal with nature, or at least with certain of its representatives, is one of the most ancient and elusive dreams of those who are disappointed by naturalism. The strange varieties of Naturphilosophie that flourished in the nineteenth century, the aesthetics of the Romantics, the current success of neo-shamanistic movements and New Age esotericism, and television's and cinema's taste for cyborgs and desiring machines—all these reactions to the moral consequences of dualism, and many others too, testify to the desire lurking within each of us, with various degrees of anxiety, to re-discover the lost innocence of a world in which plants, animals, and objects were fellow citizens. However, the Moderns' nature can emerge from its silence only by means of all-too-human intermediaries, so that no exchange, no negotiation, no contract with the host of inanimate beings is now conceivable." (Descola 2013: 397–398)

References

Albert, Jean-Pierre. 2009. "Les animaux, les hommes et l'Alliance". *L'Homme* 189: 81–114. https://doi.org/10.4000/lhomme.21997.

Argyrou, Vassos. 2005. *The Logic of Environmentalism: Anthropology, Ecology and Postcoloniality*. Studies in Environmental Anthropology and Ethnobiology. New York and Oxford: Berghahn Books.

Bird-David, Nurit. 1999. "'Animism' Revisited: Personhood, Environment, and Relational Epistemology". *Current Anthropology* 40 (S1): S67–91. https://doi.org/10.1086/200061.

Bisson, David. 2013. "Esotérisme, nature et spiritualité: Variations autour de la notion d'écologie spirituelle'". In *Nature et religions*, ed. Ludovic Bertina, Romain Carnac, Aurélien Fauches, and Mathieu Gervais, 163–72. Paris: CNRS Editions.

Brightman, Marc, and Jerome Lewis, eds. 2017. *The Anthropology of Sustainability. Beyond Development and Progress*. Houndmills, Basingstoke, Hampshire, and New York: Palgrave Macmillan. www.palgrave.com/de/book/9781137566355.

Bruckner, Pascal. 2011. *Le fanatisme de l'apocalypse: sauver la Terre, punir l'Homme*. Paris: Grasset.

Capra, Fritjof. 1997. *The Web of Life: A New Scientific Understanding of Living Systems*. New York and London: Anchor Books.

Chamel, Jean. 2016. "Visions du monde des écologistes catastrophistes: entre attente de la fin d'un monde et retrait hors du monde". In *Processus de légitimation entre politique et religion: Approches historico-culturelles et analyses de cas dans les mondes européen et extra-européen*, ed. Silvia Mancini and Raphaël Rousseleau, 281–298. Paris: Editions Beauchesne.

Chamel, Jean. 2017. "«On est tous des composts». Discours et pratiques écologistes autour des déchets organiques et des toilettes sèches". *Tsantsa* 22: 89–94.

Chamel, Jean. 2018. "«*Tout est lié*». *Ethnographie d'un réseau d'intellectuels engagés de l'écologie (France-Suisse): de l'effondrement systémique à l'écospiritualité holiste et moniste*". Doctoral thesis, Université de Lausanne.

Choné, Aurélie. 2016. "Écospiritualité". In *Guide des humanités environnementales*, ed. Aurélie Choné, Isabelle Hajek, and Philippe Hamman, 59–71. Environnement et société. Villeneuve-d'Ascq: Presses universitaires du Septentrion.

Cohn, Norman. 1957. *The Pursuit of the Millennium: Revolutionary Messianism in Medieval and Reformation Europe and Its Bearing on Modern Totalitarian Movements*. New York: Harper.

Descola, Philippe. 2005. *Par-delà nature et culture*. Paris: Gallimard.

Descola, Philippe. 2013. *Beyond Nature and Culture*. Chicago: University of Chicago Press.

Descola, Philippe. 2014a. *La composition des mondes: entretiens avec Pierre Charbonnier*. Documents et essais. Paris: Flammarion.

Descola, Philippe. 2014b. "The Difficult Art of Composing Worlds (and of Replying to Objections)". *HAU: Journal of Ethnographic Theory* 4 (3): 431–443. https://doi.org/10.14318/hau4.3.030.

Diamond, Jared M. 2005. *Collapse: How Societies Choose To Fail Or Succeed*. New York: Viking Penguin.

Dupuy, Jean Pierre. 2002. *Pour un catastrophisme éclairé: quand l'impossible est certain*. Paris: Seuil.

Faivre, Antoine. 2007. *L'ésotérisme*. Que sais-je? 1031. Paris: Presses universitaires de France.

Feuchtwang, Stephan. 2014. "Too Ontological, Too Rigid, Too Ahistorical but Magnificent". *HAU: Journal of Ethnographic Theory* 4 (3): 383–387.

Hanegraaff, Wouter J. 1998. *New Age Religion and Western Culture: Esotericism in the Mirror of Secular Thought*. Albany, NY: State University of New York Press.

Hartog, François. 2014. "L'apocalypse, une philosophie de l'histoire?" *Esprit* (6): 22–32.

Kervasdoué, Jean de. 2007. *Les prêcheurs de l'apocalypse: pour en finir avec les délires écologiques et sanitaires.* Paris: Plon.

Latour, Bruno. 1993. *We Have Never Been Modern.* Cambridge, MA: Harvard University Press.

Latour, Bruno. 2017. "Why Gaia Is Not a God of Totality". *Theory, Culture & Society* 34 (2–3): 61–81. https://doi.org/10.1177/0263276416652700.

Lovejoy, Arthur O. 1936. *The Great Chain of Being.* Cambridge, MA: Harvard University Press.

Lovelock, James E. 1986. *La Terre est un être vivant: l'hypothèse Gaïa.* L'esprit et la matière. Monaco: Le Rocher.

Lovelock, James E. 1988. *The Ages of Gaia: A Biography of Our Living Earth.* Commonwealth Fund Books. New York and London: W. W. Norton.

Lovelock, James E., and Lynn Margulis. 1974a. "Homeostatic Tendencies of the Earth's Atmosphere". *Origins of Life* 5 (1–2): 93–103. https://doi.org/10.1007/BF00927016.

Lovelock, James E., and Lynn Margulis. 1974b. "Atmospheric Homeostasis by and for the Biosphere: The Gaia Hypothesis". *Tellus* 26 (1–2): 2–10. https://doi.org/10.1111/j.2153-3490.1974.tb01946.x.

Meadows, Donella Hager, Dennis L. Meadows, Jørgen Randers, and William W. Behrens. 1972. *The Limits to Growth: A Report for the Club of Rome's Project on the Predicament of Mankind.* London: Earth Island.

Moore, Henrietta L. 2017. "What Can Sustainability Do for Anthropology?". In *The Anthropology of Sustainability*, ed. Marc Brightman and Jerome Lewis, Palgrave Studies in Anthropology of Sustainability, 67–80. New York: Palgrave Macmillan. https://doi.org/10.1057/978-1-137-56636-2_4.

Morin, Edgar. 2008. *On Complexity.* New York: Hampton Press.

Oreskes, Naomi, and Erik M. Conway. 2014. *The Collapse of Western Civilization: A View From the Future.* New York: Columbia University Press.

Rosnay, Joël de. 2014. *Le macroscope: vers une vision globale.* Points, 80. Essais. Paris: Seuil.

Roszak, Theodore, Mary E. Gomes, and Allen D. Kanner, eds. 1995. *Ecopsychology: Restoring the Earth, Healing the Mind.* San Francisco: Counterpoint.

Sahlins, Marshall. 2014. "On the Ontological Scheme of Beyond Nature and Culture". *HAU: Journal of Ethnographic Theory* 4 (1): 281–290. https://doi.org/10.14318/hau4.1.013.

Servigne, Pablo, and Raphaël Stevens. 2015. *Comment tout peut s'effondrer: petit manuel de collapsologie à l'usage des générations présentes.* Anthropocène. Paris: Seuil.

Stuckrad, Kocku von. 2005. "Western Esotericism: Towards an Integrative Model of Interpretation". *Religion* 35 (2): 78–97. https://doi.org/10.1016/j.religion.2005.07.002.

Tainter, Joseph. 1990. *The Collapse of Complex Societies*. Cambridge: Cambridge University Press.

Tan, Gillian G. 2012. "*Re-examining Human-Nonhuman Relations Among Nomads of Eastern Tibet*". Working Papers Series Two No. 38. Geelong, VIC: Deakin University. http://dro.deakin.edu.au/view/DU:30049195.

Tertrais, Bruno. 2011. *L'Apocalypse n'est pas pour demain: pour en finir avec le catastrophisme*, Médiations. Paris: Denoël.

Vullierme, Jean-Louis. 1989. *Le concept de système politique*, Politique d'aujourd'hui. Paris: Presses universitaires de France.

Worsley, Peter. 1957. *The Trumpet Shall Sound: A Study of "Cargo" Cult in Melanesia*. London: Macgibbon & Kee.

Yates-Doerr, Emily, and Annemarie Mol. 2012. "Cuts of Meat: Disentangling Western Natures-Cultures". *The Cambridge Journal of Anthropology* 30 (2): 48–64. https://doi.org/10.3167/ca.2012.300204.

This Mess Is a "World"! Environmental Diplomats in the Mud of Anthropology

Aníbal G. Arregui

In this chapter I will discuss how anthropology might crucially enact *environmental diplomacy*: a form of *comparison* and eco-political *mediation* between different ways of connecting humans and their ecosystems. In comparing the unfolding "ends of the worlds" across cultures and peoples it could help to conceptualize an idea of socio-ecological connection that requires a politically effective articulation of scientific and non-scientific rationales—and perhaps before it is too late?

In particular, this chapter will explore the eco-political approximation of a shaman and a scientist who are currently producing a common ground of communication around the future of the Amazonian rainforest. While these two spokespersons represent what might sometimes be understood as opposed ontologies or relational "worlds" (i.e., Western science vs. Amazonian shamanism), I propose to theorize the "world in-between" disclosed in their connective gestures. It will be shown how in their struggle to reach out to each other´s audiences the middle ground they

A. G. Arregui (✉)
Charles University, Prague, Czech Republic
e-mail: anibal.arregui@univie.ac.at

University of Vienna, Vienna, Austria

© The Author(s) 2019
R. Bold (ed.), *Indigenous Perceptions of the End of the World*,
Palgrave Studies in Anthropology of Sustainability,
https://doi.org/10.1007/978-3-030-13860-8_9

find—and also produce—is a messy anthropological constellation wherein different relational modes such as "naturalism" or "perspectivism" are intermingled and co-implicated. I aim at signaling here the importance of the space in between these bodies; that is, the more-than-discursive arena of negotiation around climate change as a planetary crisis and as it concerns in particular the connection between forests and rainfall patterns. May this new relational space be seen as a new "ontology" that "emerges" in the context of cosmopolitical connections around climatic crises, as Rosalyn Bold suggests (in this volume)? Persuaded that the case in focus is the bodily enactment of a world otherwise—rather than a mere discursive workout, I will stress the analytical interest of such spaces of conceptual negotiation for anthropology, and I will understand the scientist and shaman's connective gestures as an emerging form of "environmental diplomacy."

Such diplomacy will be addressed as a specific form of eco-political negotiation. The conjunctive and disjunctive effects of geo-politically and unequally distributed "global" environmental crises have made the role of certain actors, such as those I call "environmental diplomats," vital in many discussions related to the idea of "climate change." Environmental diplomats, I shall argue, are contributing a particular form of mediation between "worlds" that have become ever more mutually implicated on what is still a "shifting middle ground" (Conklin and Graham 1995).

This chapter is inspired then by the ethnographic interest in those peoples who are attempting to cooperate in order to face climate change and environmental degradation. In paying attention to specific eco-political gestures performed by the shaman and the scientist I will critically engage with the question of separating different peoples' contexts into relational "worlds," since it turns out that these "worlds" are currently being enmeshed and co-implicated by the same peoples that anthropology studies. Is this entanglement requiring to discern how indigenous peoples adapt their eco-political strategies to the stakes of global negotiations and environmental policies, as Chantalle Murtagh ethnographic case shows (in this volume)? What new conceptual tools do we need to develop in order to take seriously not only indigenous folks, but also the highly diverse "scientific" eco-logics (see Chamel, in this volume)? How do we articulate spirit and human agencies (see examples in Questa and Commanduli, in this volume), so that actually *more* "human action" becomes a cross-ontological connector to combat climate crises in this epoch we call the "Anthropocene" (see Walker and Permanto, in this volume)?

Before addressing these questions with my particular case-study in sight, it is worth noting that the co-implication of different relational worlds—that of shamanism and that of science—is not here approached as a fact but as a public enactment. However, it is my suggestion that such *mise-en-scène* neither diminishes the eco-political importance nor the empirical realism of the analysis of this particular interaction; on the contrary, it is precisely the exceptional and public quality of this dialogue that makes it anthropologically interesting and ethnographically traceable.

This chapter is divided into five sections. The first will elaborate on forms of conceiving human–environmental connections displayed by the shaman and the scientist. The second and third sections explore some of the diplomatic gestures these actors have recently made to connect eco-politically, and the different ways in which they seek to produce a common ground for negotiation. The fourth section enquires into the kind of modifications of relational limits that the shaman and the scientist enable, suggesting that their experimental and diplomatic dialogue involves a controlled form of onto-political betrayal. In the last section I will conclude by raising some thoughts on the possible roles of anthropology in mediating between those social actors that here will be identified as "environmental diplomats."

RECOGNIZING DIFFERENCE

In a TEDx talk, given in 2010 in Manaus, the Brazilian climatologist Antonio Donato Nobre explained the crucial role that the Amazonian rainforest plays in the regulation of climate on a global scale.[1] The scientist showed a variety of evocative images that illustrated the "fractal" quality of the forest and the cosmos (Nobre 2010a: 11 min into the talk), and the "invisible rivers" of air masses that flow above the South American continent (Nobre 2010a: 7 min). In resorting to such images the scientist tried to capture the attention and interest of wider audiences—the talk had nearly a million views on the Internet by the time of writing. Perhaps the more striking aspect of Nobre's strategic communication was his comparative incursion into the Yanomami shaman Davi Kopenawa's way of approaching the relationship between trees and rain, which he mobilized as follows:

I'd like to tell you a short story. Once, about four years ago, I attended a declamation, of a text by Davi Kopenawa, a wise representative of the Yanomami people, and it went more or less like this: "Doesn't the white man know that, if he destroys the forest, there will be no more rain? And that, if there's no more rain, there'll be nothing to drink, or to eat?" [...] This bugged me and I was befuddled. How could he know that? Some months later, I met him at another event and said, "Davi, how did you know that if the forest was destroyed, there'd be no more rain?" He replied: "The spirit of the forest told us." For me, this was a game changer, a radical change. I said, "Gosh! Why am I doing all this science to get to a conclusion that he already knows?" (Nobre 2010a: 14 min 50 s)

Despite the appeal to an epistemological communion of shamanism and science the scientist's claim that both sides have come to the "same conclusion" about the relationship between trees and rain can be seen as being problematic. By way of ethnographic complexification one can look, for instance, at what the shaman has specifically said about the "ecological" relation addressed:

As soon as you cut down tall trees, such as the *wari mahi* ("kapok") and the *hawari hi* ("Brazil nut"), the forest's soil becomes hard and hot. It is these big trees that make the rainwater come and keep it in the ground. The trees that white people plant, the mango trees, the coconut trees, the orange trees, and the cashew trees, they do not know how to call the rain. (Kopenawa and Albert 2013: 385)

The anthropologist Bruce Albert pointed out in an endnote to the abovequoted passage that "there is an interesting connection here with a recent theory that reveals the important effect of the tropical forest's 'pumping' atmospheric humidity has on the climate" (Kopenawa and Albert 2013: 560). The anthropologist is referring here to the Biotic Pump Theory (BPT), a new and controversial explanation about the crucial role that the Amazonian rainforest plays in atmospheric dynamics, to which Antonio Nobre has contributed extensively, conducting experiments in the Amazon (see da Rocha et al. 2009; Gorshkov and Makarieva 2007; Nobre et al. 2012; Makarieva et al. 2013).[2] Davi Kopenawa has also for his part called the scientist Antonio Nobre the Einstein of the Amazon,[3] and directly addressed white scientists or "people of ecology" in his manifesto *The Falling Sky* (Kopenawa and Albert 2013). There are then solid reasons to trace parallels between the scientist's and shaman's environmental thinking, and in particular between their respective ways of evoking the relation of trees and rain. Without any doubt, such a *mise-en-scène* of the dialogue has produced a certain eco-political "resonance," a mutual amplification of the two discourses in the face of the wider public (Prigogine and Stengers 1984: 46).

But, for all the apparent convergences, the relation between trees and rain is still significantly different. Nobre's work laying bare the physical causes and effects of micro-physical systems on climate is "worlds apart" from the work of Kopenawa in negotiating with the sky through powerful *xapiri* ("spirit beings") (Kopenawa and Albert 2013: 131–132). Furthermore, the key concepts they respectively employ to describe that relation could not differ more in their internal "extensive/intensive" logic. Kopenawa, for instance, is clear in claiming that *urihi* ("the forest") is a whole; that is, a non-fractionable entity (Kopenawa and Albert 2013: 397). Such a cosmological principle resonates also with the "intensive" and indivisible bodies of shamans as mediators between beings (Viveiros de Castro 2010). In contrast, the climatologist would in a public talk elicit a geometrical fractioning of his own body in terms of his 1/16 part of indigenous ancestry (see below for a detailed description of the event). At any rate the body of the scientist is still considered as a physical "extension" that remains therefore independent of the relation at stake. Following the perspectivist angle, two different understandings of embodied difference and transformation emerge: first, the scientist's "extensive" differences that divide space and the bodies within space into discrete units that might be isolated and then recombined for enabling change. Second, the shaman's "intensive" differences that have the potential to transform his body by an internally driven process of "becoming other," in nature and as a "whole"—rather than by its physical partition and recombination with fragments of other bodies (see Arregui 2018: 15).

Moreover, it is worth noting that the BPT suggests that the climatological relation occurring at the limits of the atmosphere–biosphere is thought of as being driven by the mechanism of a "pump." The metaphor selected seems crucial here, since it reflects plainly the naturalist aspiration to a full "mechanization of the world" (Descola 1996: 97). The relation between trees and rain is likewise one of the issues at stake for the shaman, but the "calling" of rain performed by *some* trees (Kopenawa and Albert 2013: 385) puts it not in mechanistic terms but in an animistic–relational logic where non-human beings are endowed with human agency.

It is then the very nature of ecological relations that is different: trees and rain are indeed related for both the scientist and the shaman, but there is no agreement about what such a relation consists of. In general terms the scientist is still talking about a *natural* relation based on the physical action of the forest "pumping" the atmosphere, while the shaman seems to stress the *social* agency of trees that "call" the rain. From such an oppositional perspective we seem to be confronted with two radically contrasting relational worlds. The idea that the scientist and the shaman have "come to the same conclusion" about the relation between trees and rain is in this sense hardly amenable to an anthropological standpoint, since it turns out that there is a central difference in the understanding of what "a relation" even is (Viveiros de Castro 2013: 481).

Environmental Diplomacy

It could be said that the contrast between conceptual languages leads the shaman and the scientist not only to use different vocabularies, but also to "live" in different "worlds." In this argumentative line anthropologist Peter Gow regrets in his review of *The Falling Sky*—the book in which Kopenawa has recently issued the main lines of Yanomami eco-cosmology—the (English) translator's lack of elaboration on what Kopenawa is trying to get at with the word *urihi* ("the forest"): Yanomami *urihi*, Gow (2014: 305) argues, "is not the Yanomami equivalent of our notions of nature or the environment, but rather a liveable world for Yanomami people."

Despite the importance of stressing the differences of "liveable worlds" I wish to approach the shaman's and the scientist's discourses and eco-political gestures as disclosing something other than what has been famously called "translational equivocations" (Viveiros de Castro 2004), the translative gesture by which the recognition of a radical difference or a difference of "world" becomes an end in itself. Instead, I shall compare Kopenawa's and Nobre's environmental thinking as they engage a conceptually dynamic space where differences seem to be *explicitly renegotiated* for the sake of a meaningful dialogue.

The following passage by Kopenawa might be taken as a starting point to explore such a conceptual negotiation:

> When they speak about the forest, these white people often use another word: they call it "environment." This word is also not ours and until recently we did not know it either. For us, what the white people refer to

in this way is what remains of the forest and land that were hurt by their machines […] I don't like this word. The earth cannot be split apart as if the forest were just a leftover part. We are inhabitants of this forest, and if we cut apart this way, we know that we will die with it. I would prefer the white people to talk about "nature" or "ecology" as a whole thing. (Kopenawa and Albert 2013: 397)

Kopenawa's uneasiness refers to the double meaning of the Portuguese term *meio* (middle), which is part of the Portuguese term for "environment," *meio ambiente*. In this regard, from a Yanomami perspective, *meio* suggests the violent division of the Earth "into a center—the white people's world—and a subordinate periphery—the tropical forest conceived as a residual surrounding (environment) of that center" (Kopenawa and Albert 2013: 562). For him, the rainforest is "the entire world" (Kopenawa and Albert 2013: 60) and should be "understood as a whole" (Kopenawa and Albert 2013: 397). Since the notion of *meio ambiente/* environment entails the implicit operation of "halving" the environment the shaman clearly rejects it as a translation of his indivisible *urihi*.

Yet white people's tendency to analytically dissect everything is not all that Kopenawa signals in this passage. He also stresses the potential understanding he might find in other concepts such as "nature" or "ecology." These concepts, Kopenawa argues, might be approached as "wholes," non-divisible entities that encompass all human and non-human beings and are therefore better comparable with the Yanomami *urihi*.

It is true that Kopenawa does not envisage un-making his difference with respect to white people at any point. Rather, Kopenawa engages an "imagined zone of ontological tension [that] can and should engage a form of infra-critique, gesturing at possible generative tensions, while explicitly refusing others" (Verran 2014).

> The white people who once ignored all these things are now starting to hear them a little. This is why some have invented new words to defend the forest. Now they call themselves "people of ecology" because they are worried to see their land getting increasingly hot. […] this is why we understood the new white people words as soon as they reached us. I explained them to my people and they thought: *Haixopë!* This is good! (Kopenawa and Albert 2013: 393)[4]

It is important to stress the particular struggle to establish explicit contrasts, and that these contrasts should be thinkable or understandable for an Other. I believe that such elicitation is close to what Isabelle Stengers (2005) called "diplomacy," a particular practice that seeks to allow both sides in a cosmopolitical negotiation *to be aware of* their differences as well as of the possibilities of cooperation.

Nobre's and Kopenawa's gestures are public, and in this sense escape the typical work performed behind the scenes by professional diplomats. I hence suggest approaching their art of diplomacy as being "environmental" in two senses. First, because it addresses issues that one can label as being related to the "environment." Second, in contrast to the secrecy of politico-diplomatic negotiations, the two spokespersons are willing to "air" their views (i.e., to "environmentalize" their perspectives) and thus invite their audiences to the arena of negotiation, as a form of eco-political experiment.

The main task for these two environmental diplomats is therefore the elicitation of a common ground for negotiation. In this regard their elicited contrasts are not always pursuing the confirmation of a difference per se. On diverse occasions both the scientist and the shaman have appealed to partial yet crucial forms of connectivity. These forms of connectivity are different in themselves, as what Nobre and Kopenawa might envision as "common ground" is radically different. However, the point I wish to make is not how inescapably opposed their respective cosmologies and metaphysics are. I rather aim at focusing on how the very practice of rephrasing their environmental thinking *for an Other* might be a driver of eco-political approximation. I shall situate this eco-political approximation as a tentative incursion into the conceptual imagination and the formats of communication of the Other. The following section will illustrate this observation by focusing on some public outreach gestures that the scientist and the shaman have performed in the last years.

Eliciting a (Different) Common Ground

In 2001 Nobre gave a talk about the biology and climatology of the Amazonian rainforest to an indigenous audience in Manaus. In the debate afterwards an indigenous person stood up and complained about the scientist's pretensions to know the forest. This person claimed that satellites didn't provide a right view of the forest and that only Indians could see the forest through the eyes of the spirits (cf. Nobre 2010b: 38). This is the response that Nobre offered to the indigenous audience, hoping to reduce the unexpected animosity:

I started telling them about my paternal great-grandfather, Mané Nunes, who left an indigenous tribe in Minas Gerais and married my great-grand-mother, who was black. The whiteness of my skin was the European dilution of my family, but it only came later. So, I'm proud of my native roots, and with 1/16th of indigenous blood in my veins, I felt myself to be kin of everyone in the auditorium. I went on to say that I had come in peace, and it didn't seem fair to be the subject of these attacks. (Nobre 2010b: 38–39)

Nobre also quoted Davi Kopenawa's foreword to the book *Urih A: a terra-floresta* (Albert and Milliken 2009), linking the shaman's words about the life of the Amazon rainforest to recent scientific discoveries related to the later development of the Biotic Pump Theory. Nobre suggests that his comparisons between indigenous knowledge and climate science, as well as the revelation of his 1/16 indigenous ancestry, aided in creating a better climate of conversation, since for that "small group of people the ice between the two views of the world was broken" (Nobre 2010b: 39).

The scientist's "diplomatic move" started with the elicitation of a distant but deep kinship connection with the indigenous audience. The invocation of his fraction of indigenous blood—plus another 1/16 of Afro-American lineage—was not only a way of relativizing the "whiteness of [his] skin" in the eyes of the indigenous audience, but also an analytical partition of his own body, aimed at destabilizing the scale of differences assumed by some people in the room. At least 1/8 of the scientist's body did not exactly fit, as it were, in what was assumed to be "his world." Furthermore, it was precisely this fraction of himself from which the scientist offered a new scalar frame for a negotiation of meaning. I suggest this was flagged by Nobre as a sort of "common ground" for a diplomatic encounter. Nobre's response is resonant with main-stream discourses on *mestiçagem*, based on a euro-centric conception of "blood" as the substance of ethnicity, and therefore not equivalent to Amerindian notions of "multiple natures" (Viveiros de Castro 1998) and non-substantial bodies (Vilaça 2005).

The same text is concluded by an even more provocative claim. In particular, Nobre affirms that he, like the indigenous peoples that attended that presentation in Manaus, would like to "see" the forest and the climate "with the eyes of the spirits" (Nobre 2010b: 39). One may argue that he was just addressing this audience in a metaphorical sense, or rhetorically. But it would be unfair to downplay the eco-political potential of such an affirmation. To my knowledge, it does not happen everyday that a renowned naturalist invokes "the spirits" as providers of a legitimate perspective about the world. Moreover, this comparative opening, in which science is positioned vis-à-vis indigenous knowledge, has to be framed within Nobre's project of producing a "synthesis of knowledge," aiming at confronting the climatologic future of the Amazonian region (Nobre 2014).[5] Of course, with such a gesture the scientist is conserving his own naturalistic perspective—his synthesis is still of "knowledge" and therefore not "multi-naturalistic." Yet what we consider here is the possibility of tackling two or more perspectives at the same time; that is, the disposition to take both sides "seriously" by leaving some "differences in suspension" (Candea 2011). The truly diplomatic gesture Nobre makes in fact is not the invocation of the "spirits," but the fact that he seeks a "synthesis of knowledge" which he recognizes as never fully reachable. This eco-political disposition can be detected in the scientist's wondering during his famous TED talk: "how could he know that?" (Nobre 2010a: 14 min 51 s), which suggested the recognition of his lack of knowledge of the reasons and methods by which indigenous peoples, and in particular Davi Kopenawa, know what they know.[6]

On the other side of this approximation is the shaman Davi Kopenawa. With the publication of *The Falling Sky* (Kopenawa and Albert 2013) he is not only offering a declamation of Yanomami eco-cosmology as in usual shamanic performances, he is also making it accessible to other kinds of audiences. The book can be seen as the result of the shaman's explicit interest in the disembodied forms of thinking of "whites." This is, for instance, how he compares Yanomami and white people's modes of registering and transmitting knowledge:

> This is why I want to send my words far away. They come from the spirits that stand by my side and are not copied from image skins I may have looked at. They are deep inside me. [...] Omama did not give us any books in which Teosi's words are drawn like the ones white people have. He fixed his words inside our bodies. But for the white people to hear them they must be drawn like their own, otherwise their thought remains empty. If these ancient words only come out of our mouths, they don't understand them and they instantly forget them. (Kopenawa and Albert 2013: 24)

I suggest that it is precisely because Kopenawa is able to recognize the gap of equivocation that he tries to navigate such distance by "sending [his] words far away" and fixing them in "image skins" (Kopenawa and Albert 2013: 24), in a rather unconventional way. In this regard the shaman is neither consolidating the knowledge about a difference nor undoing an anthropological distance. On the contrary, he seems to precisely aim at transforming some parameters that define the relation between "worlds," in order to dwell in them differently.

The publication of *The Falling Sky* is without any doubt a destabilizing gesture devised to transform the "world of whites" and that of the Yanomami people for the sake of the Amazonian rainforest. The whole relational constellation needs to be re-thought analytically to account for such a gesture on the part of the shaman and the anthropologist who accompanies him in this enterprise. How should we think about the effects of such "ontological debasement of the written discourse when compared with the original form of [shamanic] knowledge production and transmission (dreams, visions, accounts, and spirit-songs)" (Cesarino 2014: 290)? In other words, is it possible to grasp the anthropological importance of this gesture if we maintain the ontological opposition that the very gesture is seeking to navigate?

Viveiros de Castro has written of *The Falling Sky* that it is "a shamanic form in itself [...] in which a shaman speaks about spirits to whites, and equally about whites on the basis of spirits, and both these things through a white intermediary" (cf. Albert and Kopenawa 2013: 448). In his introduction to the Portuguese translation, Viveiros de Castro (2015) indicates the text is a "reaching-out" gesture or an attempt to overcome "cosmological monolingualism." I suggest that it is not only the content and cosmological language of the book, but also the book as disembodied materiality for communicating, that shows the diplomatic and experimental aspect of such a gesture. A book is without any doubt a technology of the whites. At the beginning of *The Falling Sky*, Kopenawa tells us that he is putting his words down on "paper skins" so that they might "penetrate the minds" of the whites (Kopenawa and Albert 2013: 12). So the original words pronounced by the shaman may not have transgressed Yanomami shamanic conventions, but their translation and their "fixation" in geometric paper skins that will travel far away reflect the shaman's intuition that the exploration of alien repertoires and formats of communication may aid to reach out to the imagination of Others.

The text written with Albert is not the first attempt Kopenawa made at bringing his message beyond the context of the Yanomami people.

In December 2000 in Kopenawa's community (in the Brazilian state of Amazonas) the shaman held up, in the context of a shamanic ritual, a satellite image of the Earth's extension, the image of a planisphere (see Albert 2014). Amerindian shamanic knowledge originates in a shaman's personal and even intimate contact with the "images" of what exists. These images are not the existing beings in themselves, but spirit-entities that reveal deeper qualities of beings, and that only shamans can "bring down" and see (Kopenawa and Albert 2013: 29). In this regard the use of a satellite image that everybody could see directly represented the disembodying of part of shamanic knowledge, a gesture that therefore destabilized the conventional format of Yanomami shamanic practices.

According to Bruce Albert the picture of that event, taken by a French friend, instantiated a "sort of intercultural *mise en abyme*" (Albert 2014: 237). Albert suggests that in holding that satellite image, Kopenawa

> is displaying an innovative form of political shamanism that marked the beginning, for the Yanomami, of a new kind of "war of images." Along with the incursion of their goods and diseases, which had begun back in the 1970s, the white people's images had been gradually infiltrating into the Yanomami forest. What the shamans were thus seeking, in this instance, was to capture and turn the power of those images around in order to keep their creators from advancing over their land. Taken in solidarity with this resistance movement, Hervé's photograph in fact produces another image of the confrontation between shamanic imagination and Western imagery, but from a perspective—and for an audience—rooted in the world of its origins, our world. (Albert 2014: 237–238)

Gestures like the one just described address not only the Yanomami people but also are obviously performed for outsiders. These gestures need to be approached with analytical tools that account for how the gap between different relational "worlds" is being navigated in the exercise of eco-political mediations. In the next section I make some conceptual considerations for attending to those actors whose mediations involve also risky outreaching gestures.

Diplomatic Betrayal

I wish to draw attention to the fact that the shaman and the scientist may be seen as spokespeople who conduct a sort of environmental diplomacy; and I understand this as an exercise of cross-cultural translation that may aid us to approach positions within a given eco-political nego-tiation. However, the shaman and the scientist are not engaged in direct negotiation, occurring behind the scenes, as in conventional diplomatic practices. It is rather a communication strategy that consists in making public the many points their respective worlds have (or might have) in common. It is worth recalling that Nobre's and Kopenawa's gestures are pushed by the urgent need for cooperation to combat ongoing processes of environmental degradation. As Latour suggests:

> The multiplication of such diplomatic scenes will become even more important when the acceleration of ecological mutations forces the inhab-itants of the shrinking domains of life into finding out how to compose the common world that they are supposed to inhabit, if not peacefully, at least without exterminating one another. (Latour 2014)

Indeed, what the great majority of Western scientists would iden-tify as "climate change" constitutes a very diverse arena for negotia-tion between different peoples and beings. The question at stake here is: What makes the actors of a staged eco-political dialogue into "diplo-mats"? I argue that they are putting in practice an environmental diplo-macy because they do not simply aspire to modify their interlocutors' line of action by imposing a one-sided version of a problem; they also show a disposition to consider a given situation from the perspective of their interlocutor. It seems that both the scientist and the shaman explic-itly and respectfully acknowledge their counterpart´s basic principles of human–environment relationality (i.e., animism and naturalism). The gestures described might be seen as scenic, yet they also set up a context for dialogue where differences are made explicit and respected.

Furthermore, as we know diplomatic work always involves the assumption of a risk, which derives from the fact that diplomats might be, as Stengers (2005: 193) would have it, "denounced as traitors when arriving back home." Both the scientist and the shaman have proven their ability to betray conventions and expand their fields of expertise to open them to non-conventional ways of criticism and dialogue. On the side of the scientist, it is worth noting that the Biotic Pump Theory (BPT) has since its inception sparked intense controversy among atmos-pheric scientists (Gorshkov and Makarieva 2007; Makarieva et al. 2013). The originality of the BPT makes it indeed very attractive to the wider public, while it is a focus of intense discussions and controversies among scientists of the field, as noted by the editor of a leading journal of the field, the *British Medical Journal* in an editorial.[7]

As regards the shaman, he has declared himself aware that his interest and openness to the world of whites is being used to accuse him of having "become a white man" and "not being a real Yanomami" (Kopenawa and Albert 2013: 436). The shaman was first educated by Christian pastors, and a prevalent theme throughout his biography is that of a return, "becoming a shaman": a vital path that allowed him to reinforce his central role in mediation between the Yanomami people and outsiders.

Betrayal and diplomacy seem to be closely attached to the practice of eco-politics. I would then suggest, paraphrasing Viveiros de Castro's (2004) famous essay, that environmental diplomats conduct a "controlled" form of betrayal. Rather than constituting a sort of ontological rupture the controlled betrayal would involve the minimum dose of treason toward one's own relational principles that is necessary to turn an "equivocation" into an eco-politically productive communication. The controlled betrayal is in other words a diplomatic form of "cosmopolitical" translation: one that explicitly—and within particular ethnographic situations—attends to how different perspectives in contact can affect each other in productive ways, stressing their differences and equivocations as well as the logics of their mutual reinforcement (Blaser 2016).

CODA: THE SHAMAN'S SHAMAN

This chapter has highlighted the efforts that a scientist and a shaman are making to stress the need to protect the rainforest. In particular, I have drawn attention to the fact that they refer to each other's perspectives on the connection between forest trees and rainfall—not as opposed relational worlds, but as complementary and co-implicated eco-political positionings. In their search for publicity and their communicative experimentation it becomes clear that both actors aim to reach out to and make their "worlds" thinkable for "alien" audiences. I have argued that due to their connective, public gestures these actors are not only spokespersons but also "environmental diplomats" who pursue a coordinated strategy against the despoiling of Amazonian ecosystems.

Despite both the scientist and the shaman being highly respected in their fields it is clear that their public life brings them beyond the limits of their usual relational constellations. The targeting of different audiences and experimentation with unconventional forms of communication are part of these actors' eco-political strategies. The liminal work carried out by environmental diplomats puts them and the anthropologists who may analyze their work in an inevitably messy arena. First, the very elements of this arena are not easily discerned. As I show elsewhere (Arregui 2018), for Nobre and Kopenawa relational principles or perspectives are eco-politically "embodied" and mimetically co-implicated. Within this scenic dialogue the shaman offers disembodied forms of communication—a book, a map—following white people's customs. From the other side a climatologist speaks about the ecology of the rainforest in making references to embodied forms of knowledge and emphasizing how masterfully are these forms of knowledge used by Amazonian shamans (Arregui 2018: 15–20). In their connective gestures the shaman's and the scientist's "perspectival" and "naturalistic" relational logics become complexly enmeshed, publicly enacted from both sides at the same time.

Nobre and Kopenawa enact "worlds" where perspectivism and naturalism are co-implicated in different ways, in the sense of one relational mode becoming complexly implicated in the other. One of the problems of the ontological turn, and perhaps the reason for some critiques of it, is that it has not yet sufficiently explored the extent to which different "worlds" are characterized by these kinds of mimesis and co-implication (i.e., by the messy relational space to which worlds tend to converge; Viveiros de Castro 2013: 482).

In Amazonian anthropology, naturalism and perspectivism have been respectively related—and mutually opposed—to the metaphysical notions of extensive and intensive differences (see Viveiros de Castro 2010). But these metaphysical principles are also to be understood as complexly enmeshed. Such co-implication is not only a metaphysical critique that Deleuze posed to Bergson's too static separation of the extensive—as differences of degree—and the intensive—as differences of nature (Deleuze 1968: 308; see also De Landa 2005), but there is also an anthropological problem at stake here.

The main issue is that the language "world," as used in anthropology, has often emphasized different contexts as being contained in self-evident worlds that exist and that should be explained just in their own terms (Pina-Cabral 2014). Despite the insistence in the relation per difference the whole space of co-implication thus runs the risk of being downplayed, erased from the analytical landscape. This has fostered very inspiring and sophisticated analytical tools for recasting what can be seen as internal, indigenous sociologies. But it seems a limited, or rather a limiting solution, when it comes to approaching what these same indigenous actors are doing to confront, and even re-frame for their own benefit, current global economic and ecological issues (see Turner 2007).

The context of climate change is particularly challenging for all kinds of anthropological theories. If one assumes the legitimate task of de-naturalizing the environment to see it as a "society of societies" (i.e., a truly "international arena"; Danowski and Viveiros de Castro 2015), then it becomes clear very soon that at this level of socio-ecological interaction there are many kinds of difference and connection between perspectivism and naturalism. Ecological crises constitute therefore an analytical space that needs to be further explored to re-establish the equilibrium between the ontologically committed and ethnographically actualized accounts of how different peoples are trying to cooperate to face the "end of the(ir) world."

In such a relational situation the role of the anthropologist may appear as that of an environmental or rather eco-political mediator. With healthy irony, Roy Wagner points out in his review of *The Falling Sky* that the role of Bruce Albert, as anthropologist, is that of a "universal mediator of voices" which in this particular case makes him appear as "the shaman's shaman" (Wagner 2014: 297). However, the question of what exactly it means to "mediate" between voices or "worlds" is a serious one and remains still unanswered. In the case we are considering, the ironic characterization of the ethnographer as a "shaman's shaman" could be grandiloquently extended to depict us—anthropologists and the like—as the "scientists' scientists," or even as the "diplomats' diplomats." At any rate the main question to be raised in the light of the case presented above is perhaps what should anthropology do to mediate among scientists, shamans, and diplomats without restraining the role that each of these actors is having in the human coping with current ecological crises.

Despite the urgent actuality of the problem at stake—climate change as the end of the world—in this chapter I have tried to address an old question that has dogged our discipline from the onset. Recounting how he abandoned philosophy to "become an ethnographer," Lévi-Strauss (1955: 53) already noted the intellectual impoverishment he saw in taking metaphysical "worlds" as the substance of a merely logical *gymnastique*. Much was said afterwards, but the ontological superimposition of the anthropologist's Self and the world of the Other—even if it is for the sake of decolonizing our thought—still takes the risk of practicing a kind of "suprarationalism" where the first resigns precisely to his ethnographic sensibility (Lévi-Strauss 1955: 61). What it is to practice "ethnography" is surely another theme for discussion, and this chapter is not an example of its more orthodox interpretation. However, I have tried to transparently describe some very specific, empirical situations in which a scientist and a shaman publicly display the possibilities of a common eco-political agenda, the positive side of seeing the actual entanglements of ethno-metaphysical differences, and how this may stimulate very different audiences to reconsider the ecological importance of the Amazonian rainforest.[8] And it is clear that beyond mutually exclusive cosmological principles, attention to the specific interactions of different actors does not disclose neat lines of separation. On the contrary, a messy world of discussion seems to be the natural niche of diplomats. Their diplomacy is not limited to the mere evocation of cosmological differences, but thrives in a conceptual mud in which they seem to need to hold each other's hands to ensure a minimum equilibrium. As for the mediating figure of the anthropologist, it could be said that metaphysical workouts have proven to be a very healthy practice, yet it is perhaps time for the shaman's shamans to get again into the mud and tell us how to analytically dwell in this mess our friends and informants are currently dealing with.

Notes

1. https://www.ted.com/talks/antonio_donato_nobre_the_magic_of_the_amazon_a_river_that_flows_invisibly_all_around_us. Accessed 8 May 2017).
2. Unlike the widely accepted theory that holds that it is the Earth's rotation rate, the heating of the oceans, and atmospheric depth that are the main drivers of air masses, the BPT suggests that a more powerful driver of winds might be found in the biosphere, in particular in tropical forests. According to the BPT the condensation of water vapor released by trees creates a rapid depression—a low-pressure zone—that powerfully attracts moist, high-pressure masses of air, which can be "pumped" along thousands of kilometers, from oceans and seas toward the forest.
3. http://terramagazine.terra.com.br/interna/0,,OI5437416-EI16863,00.html. Accessed 7 July 2015.
4. If "ecology" is one of these words perceived by Kopenawa as "good to think with" cross-environmentally, the Anthropocene might also be a concept that can trigger meaningful discussions among Western scientists and indigenous spokespeople. As an example see the cross-disciplinary and

cross-cultural gathering The Thousand Names of Gaia (Rio de Janeiro, September 2014). See https://thethousandnamesofgaia.wordpress.com/program/. Accessed 23 July 2018.

5. In this report Antonio Nobre not only praises the "holistic approach" of such naturalists as Alexander von Humboldt, who was the first to suggest a connection between forests, air humidity, and climate (Nobre 2014), but he also concludes by summoning the indigenous knowledge that may show us the "many excellent alternatives for reviving the respectful (and technological) coexistence that ancient civilizations once enjoyed with the forest" (Nobre 2014: 36).

6. I have argued that in this case both the scientist and the shaman do not only recognize the equivocations but also experiment with "mimicries" that are devised to precisely navigate (not undo) these differences eco-politically (see Arregui 2018: 15–20).

7. http://blogs.bmj.com/bmj/2013/01/28/richard-smith-the-editor-thinks-your-paper-is-nonsense-but-will-publish-anyway/.

8. While it is difficult to assess the effect in the audiences it seems clear that the cross-cultural reference to other forms of knowledge has increased the potential impact of the ecological thinking of the shaman and the climatologist. *The Falling Sky* is being translated into several languages (e.g., French original, English, Portuguese, Italian) and is receiving important publicity on the part of big indigenous rights organizations such as Survival International (https://www.youtube.com/watch?v=G_lBxXrX-Eis). As for the climatologist, Antonio Nobre's TEDx talk has had as of July 2018 more than one million views on the internet (https://www.ted.com/talks/antonio_donato_nobre_the_magic_of_the_amazon_a_river_that_flows_invisibly_all_around_us).

REFERENCES

Albert, Bruce. 2014. *Yanomami: Back to the Image(s)*, 237–248. Paris: Fondation Cartier.

Albert, Bruce and William Millken. 2009. *Urihi A: A Terra-Floresta Yanomami*. São Paulo: Instituto Socioambiental.

Arregui, Anibal. 2018. "Embodying Equivocations: Ecopolitical Mimicries of Climate Science and Shamanism." *Anthroppological Theory*. https://doi.org/10.1177/1463499617753335.

Blaser, Mario. 2016. "Is Another Cosmopolitics Possible?" *Cultural Anthropology* 31 (4): 545–570.

Candea, Matei. 2011. "Endo/Exo." *Common Knowledge* 17: 146–150.

Cesarino, Pedro de N. 2014. "Ontological Conflicts and Shamanistic Speculations in Davi Kopenawa's *The Falling Sky*." *HAU: Journal of Ethnographic Theory* 4 (2): 289–295.

Concklin Beth A., and Graham Laura. 1995. "The Shifting Middle Ground: Amazonian Indians and Eco-Politics." *American Anthropologist New Series* 97 (4): 695–710.

Danowski, Debora and Eduardo Viveiros de Castro. 2015. "Is There Any World to Come?" Available http://supercommunity.e-flux.com/authors/deborah-danowski/. Accessed 12 January 2016.

da Rocha, H. R., et al. 2009. "Patterns of Water and Heat Flux Across a Biome Gradient from Tropical Forest to Savanna in Brazil." *Journal of Geophysical Research-Biogeosciences* 114: G00B12. https://doi.org/10.1029/2007JG000640.

De Landa, Manuel. 2005. "Space: Extensive and Intensive, Actual and Virtual." In D*eleuze and Space*, edited by I. Buchannan and G. Lambert, 80–88. Edinburgh: Edinburg University Press.

Deleuze, Gilles. 1968. *Différence et Répétition*. Paris: Presses Universitaires de France.

Descola, Philippe. 1996. "Constructing Natures: Symbolic Ecology and Social Practice." In *Nature and Society: Anthropological Perspectives*, edited by P. Descola and G. Pálsson, 82–102. London: Routledge.

Gorshkov, Viktor. G., Makarieva Anastassia. 2007. "Biotic Pump of Atmospheric Moisture as Driver of the Hydrological Cycle on Land." *Hydrology and Earth System Sciences* 11: 1013–1033.

Gow, Peter. 2014. "'Listen to Me, Listen to Me, Listen to Me, Listen to Me ... 'A Brief Commentary on *The Falling Sky* by Davi Kopenawa and Bruce Albert." *HAU: Journal of Ethnographic Theory* 4 (2): 301–309.

Kopenawa, Davi and Bruce Albert. 2013. *The Falling Sky: Words of a Yanomami Shaman*. Cambridge, MA: The Belknap Press of Harvard University Press.

Kuper, Adam. 2003. The Return of the Native. *Current Anthropology* 44 (3): 389–402.

Latour, Bruno. 2014. "Another Way to Compose the Common World." *HAU: Journal of Ethnographic Theory* 4 (1): 301–307.

Lévi-Strauss, Claude. 1955. *Tristes Tropiques*. Paris: Plon.

Makarieva, Anastassia, et al. 2013. Where Do Winds Come from? A New Theory on How Water Vapor Condensation Influences Atmospheric Pressure and Dynamics. *Atmospheric Chemistry and Physics* 13: 103956.

Nobre, Antonio D. 2010a. The Magic of the Amazon: A River That Flows Invisibly All Around Us. TED Talk (Video and Transcript). Available https://www.ted.com/talks/antonio_donato_nobre_the_magic_of_the_amazon_a_river_that_flows_invisibly_all_around_us/transcript. Accessed on 8 April 2015.

Nobre, Antonio. 2010b. "Floresta e Clima. Saber Indígena e Ciência." In *Manejo do mundo: conhecimentos e práticas dos povos indígenas do Rio Negro, Noroeste amazônico* (Org.) A. Cabalzar, 38–45. Brasil: Instituto Socioambiental & Federação das Organizações Indígenas do Rio Negro.

Nobre, Antonio. 2014. *The Future Climate of Amazonia*. Sao Paulo: Edition ARA, CCST-INPE e INPA.

Nobre, Antonio D., et al. 2012. "Distributed Hydrological Modeling of a Micro-Scale Rainforest Watershed in Amazonia: Model Evaluation and Advances in Calibration Using the New HAND Terrain Model." *Journal of Hydrology* 462: 15–27.

Pina-Cabral, João de. 2014. "World." *HAU: Journal of Ethnographic Theory* 4 (1): 49–73.

Prigogine, Ilya, and Isabelle Stengers. 1984. *Order Out of Chaos: Man's New Dialogue with Nature*. New York: Bantam Books.

Stengers, Isabelle. 2005. "Introductory Notes to an Ecology of Practice." *Cultural Studies Review* 11 (1): 183–196.

Turner, Terence S. 2007. "Indigenous Resurgence, Anthropological Theory, and the Cunning of History." *Focaal—European Journal of Anthropology* 49: 118–123.

Verran, Hellen. 2014. "Anthropology as Ontology Is Comparison as Ontology." *Theorizing the Contemporary, Cultural Anthropology Website*. https://culanth. org/fieldsights/468-anthropology-as-ontology-is-comparison-as-ontology.

Vilaca, Aparecida. 2005. "Chronically Unstable Bodies: Reflections on Amazonian Corporalities." *Journal of the Royal Anthropological* 11: 445–464.

Viveiros de Castro, Eduardo. 1998. "Cosmological Deixis and Amerindian Perspectivism." *The Journal of the Royal Anthropological Institute* 4 (3): 469–488.

Viveiros de Castro, Eduardo. 2004. "Perspectival Anthropology and the Method of Controlled Equivocation." *Tipití: Journal of the Society of Lowland South-America* 2 (1): 3–22.

Viveiros de Castro, Eduardo. 2010. "Intensive Filiation and Demonic Aliance." In *Deleuzian Intersections: Science, Technology and Anthropology*, edited by C. B. Jensen and K. Rodje, 219–254. Oxford: Berghahn Books.

Viveiros de Castro, Eduardo. 2013. "The Relative Native." *HAU: Journal of Ethnographic Theory* 3 (3): 473–502.

Viveiros de Castro, Eduardo. 2015. "Prefacio. O Recado da Mata." In *A queda do céu. Palavras dum xama Yanomami*, edited by B. Albert and D. Kopenawa, 11–42. Sao Paulo: Companhia das Letras.

Wagner, Roy. 2014. "The Rising Ground." *HAU: Journal of Ethnographic Theory* 4 (2): 297–300.

Epilogue: Indigenous Worlds and Planetary Futures

Bronislaw Szerszynski

What does it mean to talk of "the end of the world"? "World" is a polyvalent word, unique to the Germanic languages, that in the history of its evolving meaning holds together in creative tension the human and the planetary, the spatial and the temporal, and the idea of human life as an independent given and that of it as always existing in relation to supernatural realities. But what if we took the word "world" to mean "planet"? According to modern cosmology the world as a planet is 4.5 billion years old and will probably last about another 5 billion before it is burnt up by an aging, expanding sun. Such a sentence only seems to make sense in the context of a particular "episteme," involving interconnected ideas such as space-time, atoms, energy, galaxies, stars, and planets, themselves supported by a vast interconnected set of knowledge practices, instruments, and devices. Here the richness of the idea of a "world" seems to have been lost; we seem to be a particularly long way from the "worlds" of Amerindian peoples; and, in a literal sense, we seem to be a long way from the *end* of the world.

B. Szerszynski (✉)
Department of Sociology, Lancaster University, Lancaster, UK
e-mail: bron@lancaster.ac.uk

© The Author(s) 2019
R. Bold (ed.), *Indigenous Perceptions of the End of the World*,
Palgrave Studies in Anthropology of Sustainability,
https://doi.org/10.1007/978-3-030-13860-8_10

203

But there are aspects of our emerging understanding of planets that, I would argue, have a curious affinity with Amerindian thought. Planets are dense, folded spaces of mutuality, which still retain the imprint of their emergence from the immanence of the molecular cloud out of which the planetary system of which they are a part took form. As they coalesce into materially closed assemblages circling in the void, they differentiate internally into different compartments and entities—which parts then come into close association in ways that shape and define and in a way "sense" each other (Margulis 1998; Clarke and Hansen 2009). In these dense, folded spaces, phenomena such as life (Dupré 2012), energy (Barry 2015), and geochemical processes are highly relational and processual phenomena through which apparently separate entities are connected in webs of interrelationship and transformation. And multiple temporalities are generated within the extended body of the planet, so that multiple world-beginnings and world-endings are internal to the process of planetary self-organization, rather than singular bookends to a unitary planetary story.

Although the empirical sciences of the Earth have played a huge role in shaping such insights, it sometimes feels as if Western scientific metaphysics is not in fact the best framework for articulating them—as if a deeper understanding of planetarity is struggling to get out of the straitjacket of Western naturalism. Indigenous ontologies—and not least the Amerindian and related cosmologies that are foregrounded in this book—might seem a more hospitable context for the features of planetary being listed above. This is of course not a wholly novel claim, but the chapters in this volume give new depth to that intuition. Such cosmologies foreground fluidity and transformation, as different entities shift between what might otherwise be seen as wholly separate categories; multiple times seem to operate at once, and originary immanence is just a blink away; beings exist in and are defined by meshworks of reciprocity and generosity; grasping the deepest truth can involve departing radically from everyday perception and knowledge; everything is alive, aware, a potential interlocutor. In ways that are resonant with contemporary theory, the chapters suggest that the Earth is not a dead mechanism, but sensitive (Latour 2017), ticklish (Stengers 2015), and maybe dangerous (Hamilton 2017). And such cosmologies are grounded in forms of social metabolism that resonate with them: that involve passing on the "accursed share" of excess production in moments of gift and festival, rather than reinvesting it in endless, industrial growth

(Bataille 1988), and that resist the accelerating linear flows caused by tapping into fossil fuel and mineral deposits and thereby unleashing the gargantuan chemical energy gradients that lie between different strata of the Earth.

In an important paper the philosopher Jane Howarth (1995) insists that the human tendency to make meaningful connections between the "natural environment" and our inner "mood" (by which she means affect without a clear object) can richly repay philosophical attention. To make that connection is not simply to make a causal claim: that the weather affects mood. Neither is it simply an anthropomorphism, a pathetic fallacy, a category mistake, a projection of capacities such as sentience, agency, or emotions onto non-human or even non-biological entities that (according to Western thought) do not and cannot possess those capacities. What are we doing, Howarth asks, when we say things like "the angry sea rages, thunders, is turbulent, frenzied, destructive, forceful, dashes against the cliffs" (Howarth 1995: 115)? Her answer is that part of learning what particular moods are involves experiencing the atmospheres of the natural world that correspond with them. The implication of this line of thought is that perhaps humans do not start with a fully formed internal emotional world, which they then might project outward onto the blank canvas of nature; it may be more true to say that humans-to-be arrive in the midst of a more-than-human world that already possesses different emotional tones—and that humans-to-be have to learn to *introject* these to structure and articulate their own inner experience, and thus become fully human.

The indigenous voices here enrich this insight beautifully. The peoples that inhabit these pages feel themselves to be open to forces in the environment around them. Teresa Brennan argues that the modern idea of psychic self-closure is inextricably linked with capitalist modernity, with its proliferation of commodities and technological domination of nature. In an echo of Latour's (1993) argument that, as modernity tries to separate and purify nature and culture, hybrids actually proliferate, she suggests that the notion of psychic closure from wider energetic flows has helped to drive the breathtaking energetic profligacy of modern society, and the incessant conversion of the energies of life into dead commodities (Brennan 2000). By contrast, accounts such as that of Bold (chapters "Introduction: Creating a Cosmopolitics of Climate Change" and "Contamination, Climate Change, and Cosmopolitical Resonance in Kaata, Bolivia" in this volume) of how the air in the Andes is experienced by its peoples as suffused with moods, with animate and living presences, suggests a very different eco-politics of "vibrant matter" (Bennett 2010).

What does the end of a world look like? A world can be seen as a system of signs, and a system of signs can be disrupted and deranged. Gregory Bateson argued that, because ecological order is semiotic, ecological collapse starts with a disruption in the communicative order of an ecosystem, which only later becomes manifest in physical changes (Bateson 1972). We see that in the chapters here—that the semiotic relations between different beings together are shifting, no longer always make sense, and often seem to be moving away from liveliness and animacy. Comandulli, for example (chapter "A Territory to Sustain the World(s): From Local Awareness and Practice to the Global Crisis" in this volume), reports that the Ashaninka of Peru and Brazil can no longer trust the signs in the world around them on which they depend for their livelihoods. But if worlds are semiotic it does not mean that they are merely a set of signs to be perceived. This book reminds us that a world is also a set of relations, obligations, and mutual dependencies; the indigenous voices in these chapters warn what happens when these relationships are not maintained. For the Masewal discussed by Questa (chapter "Broken Pillars of the Sky: Masewal Actions and Reflections on Modernity, Spirits, and a Damaged World" in this volume), mountains are living, active entities or collections of entities, sentient "reservoirs of life" that can repopulate the world with new beings—but are being killed by megaprojects. In a comparable way the Q'eqchi' Maya people in Permanto's chapter ("The End of Days: Climate Change, Mythistory, and Cosmological Notions of Regeneration" in this volume) speak of the *tzuultaq'as*, lords of the hills that dwell in hills and mountains, with whom an "existential reciprocity" must be maintained—but say that the deforestation of a hill indicates that the *tzuultaq'a* of that particular hill is disempowered, or even dead.

The Anthropocene threatens to be an epoch of planetary forgetting (Szerszynski 2019), a loss not just of genetic, ecological, and climate memory but also of cultural memory, as meaningful local places and their multi-species forms of life are deterritorialized into globalized non-places, defined only by their location in flows of resources and commodities. The chapters of this book give us many depressing examples of these "ends of the world." But more, these chapters show us that indigenous cultures give us a different way of thinking about *what forgetting is*. Unlike ontologically lazy Westerners, who believe that the world arrives already as a unified, composed whole (Latour 2017: 86) and that it will go on existing whatever they do or do not do (Law 2002), we see that Amerindian peoples know that the world has to be actively maintained in being, through gift exchanges among humans and non-humans, and that the apocalypse is an ever-present danger. For such peoples in forgetting thus lies not just an epistemological tragedy but an *ontological tragedy*; they recognize that for the planet to endure in its power to go on, and its power of natality (Arendt 1958)—the power to produce new beginnings as well as new endings—its memory systems have to be actively maintained. This means not just maintaining traditional knowledge, but also the subsistence practices and traditional lifeways that make such knowledge intelligible (Moeller 2018).

Yet seeing the peoples of Latin America merely as custodians of an invaluable heritage would be an error. Like the wider movement of activists and researchers seeking to preserve and record biodiversity, to protect cultural diversity and political freedoms, to deepen our knowledge about environmental change, and to prevent the erosion of regulatory powers, indigenous peoples are not just building a museum for the *actual*. They are protecting the *virtual*—defending important, hard-won planetary preconditions that enable the future to arrive. As Danowski and Viveiros de Castro (2016: 123) put it, "Amerindian collectives, with their comparatively modest populations, their relatively simple technologies that are nonetheless open to high-intensity syncretic assemblages, are a 'figuration of the future,' not a remnant of the past." Those of us, anthropologists, activists and others, who see in indigenous thought and practice a vital element in the task of making the Earth's future otherwise have to attend to the important issues that Arreguí (chapter "This Mess Is a "World"! Environmental Diplomats in the Mud of Anthropology" in this volume) raises about environmental diplomacy and cosmopolitical translation as a "controlled form of betrayal." In making different ways of knowing speak to each other, in convening the "people of Gaia" (Latour 2017), the threat of treachery and apostasy can never wholly be kept at bay.

But the messy task must continue. Indeed, I have argued (Szerszynski 2017) that we in the industrialized West have no longer any choice but to take notice of the "earth beings" and other spiritual agencies reported in these pages and elsewhere. To paraphrase William Gibson's famous phrase about the future, "the apocalypse is already here—it's just not evenly distributed." Or as Evan Calder Williams puts it in his

Combined and Uneven Apocalypse (2010), "[t]he world is already apocalyptic. Just not all at the same time." This is a recognition that the end of the world has already happened for some (Danowski and Viveiros de Castro 2016)—but also that the "combined and uneven" nature of the Anthropocene earth extends to the spiritual realm. Let me expand on this. Trotsky's (1932) concept of "combined and uneven development" itself combined Lenin's recognition that different societies progress at different speeds with his own insight that this unevenness was itself produced and maintained by imperialist economic relations. Spatial variegation is an inherent rather than accidental feature of the capitalist world system (Jessop 2011); it is not the case that different economies on the same planet can independently follow the same path and all come to enjoy the same benefits of advanced capitalism. If we abandon the linearity implied in Trotsky's ideas of stages, and of development as fast and slow, and instead see the "unevenness" as about the coming into correspondence of different ontologies, we might call the Anthropocene a "combined and uneven geo-spiritual formation." The combination here implies we can no longer keep these other naturecultures at a distance—their agencies, stirred and uprooted by the industrial mobilization of matter across the planet, are starting to haunt the subconscious of modernity, requiring new kinds of response (Szerszynski 2017). In such a context books like *Climate Change as the End of the World* should become required reading.

References

Arendt, Hannah. 1958. *The Human Condition*. Chicago: University of Chicago Press.

Barry, Andrew. 2015. "Thermodynamics, Matter, Politics." *Distinktion: Journal of Social Theory* 16 (1): 110–125.

Bataille, Georges. 1988. *The Accursed Share: An Essay on General Economy*, vols. 1–3. New York: Zone Books.

Bateson, Gregory. 1972. *Steps to an Ecology of Mind: Collected Essays in Anthropology, Psychiatry, Evolution, and Epistemology*. Aylesbury: Intertext.

Bennett, Jane. 2010. *Vibrant Matter: A Political Ecology of Things*. Durham: Duke University Press.

Brennan, Teresa. 2000. *Exhausting Modernity: Grounds for a New Economy*. London: Routledge.

Clarke, Bruce, and Mark B.N. Hansen, eds. 2009. *Emergence and Embodiment: New Essays on Second-Order Systems Theory*. Durham, NC: Duke University Press.

Danowski, Déborah, and Eduardo Viveiros de Castro. 2016. *The Ends of the World*. Oxford: Polity.

Dupré, John. 2012. *Processes of Life: Essays in the Philosophy of Biology*. Oxford: Oxford University Press.

Hamilton, Clive. 2017. *Defiant Earth: The Fate of Humans in the Anthropocene*. Cambridge: Polity.

Howarth, Jane. 1995. "Nature's Moods." *British Journal of Aesthetics* 35 (2): 108–120.

Jessop, Bob. 2011. "Rethinking the Diversity of Capitalism: Varieties of Capitalism, Variegated Capitalism, and the World Market." In *Capitalist Diversity and Diversity Within Capitalism*, edited by Geoffrey T. Wood and Christel Lane, 209–237. London: Routledge.

Latour, Bruno. 1993. *We Have Never Been Modern*. Translated by Catherine Porter. Hemel Hempstead: Harvester Wheatsheaf.

Latour, Bruno. 2017. *Facing Gaia: Eight Lectures on the New Climatic Regime*. Cambridge, UK and Medford, MA: Polity Press. https://ebookcentral.proquest.com/lib/duke/detail.action?docID=4926426.

Law, John. 2002. *Aircraft Stories: Decentering the Object in Technoscience*. Durham, NC: D uke University Press.

Margulis, Lynn. 1998. *The Symbiotic Planet: A New Look at Evolution*. London: Weidenfeld & Nicolson.

Moeller, Nina Isabella. 2018. "Plants That Speak and Institutions That Don't Listen: Notes on the Protection of Traditional Knowledge." In *Food Sovereignty, Agroecology and Biocultural Diversity: Constructing and Contesting Knowledge*, edited by Michel P. Pimbert, 202–233. London: Routledge.

Stengers, Isabelle. 2015. *In Catastrophic Times: Resisting the Coming Barbarism*. Translated by A. Goffey. Ann Arbor, MI: Open Humanities Press.

Szerszynski, Bronislaw. 2017. "Gods of the Anthropocene: Geo-Spiritual Formations in the Earth's New Epoch." *Theory, Culture & Society* 34 (2–3): 253–275.

Szerszynski, Bronislaw. 2019. "How the Earth Remembers and Forgets." In *Political Geology: Active Stratigraphies and the Making of Life*, edited by Adam Bobbette and Amy Donovan, 219–236. London: Palgrave Macmillan.

Trotsky, Leon. 1932. *The History of the Russian Revolution*. Translated by Max Eastman. London: Victor Gollancz.

Williams, Evan Calder. 2010. *Combined and Uneven Apocalypse*. Winchester: Zero Books.

INDEX

The manufacturer's authorised representative in the EU is Springer
Nature Customer Service Centre GmbH, Europaplatz 3, 69115 Heidelberg,
Germany. If you have any concerns regarding our products, please
contact ProductSafety@springernature.com

Printed and bound by CPI Group (UK) Ltd, Croydon, CR0 4YY
23/04/2026
02095601-0004